Model Predictive Control on Open Water Systems

Model Predictive Control on Open Water Systems

Proefschrift

ter verkrijging van de graad van doctor
aan de Technische Universiteit Delft,
op gezag van de Rector Magnificus prof. dr. ir. J.T. Fokkema,
voorzitter van het College voor Promoties,
in het openbaar te verdedigen op maandag 26 juni 2006 om 12:30 uur

door Peter-Jules VAN OVERLOOP
werktuigkundig ingenieur
geboren te Bergen op Zoom

Dit proefschrift is goedgekeurd door de promotoren:
Prof. ir. R. Brouwer
Prof. ir. O.H. Bosgra

Samenstelling promotiecommisie:

Rector Magnificus, voorzitter
Prof. ir. R. Brouwer, Technische Universiteit Delft, promotor
Prof. ir. O.H. Bosgra, Technische Universiteit Delft, promotor
Dr. S. Dijkstra, Technische Universiteit Delft
Prof. dr. ir. N.C. van de Giesen, Technische Universiteit Delft
Dr. A.J. Clemmens, United States Department of Agriculture
Prof. dr. ir. A.C.P.M. Backx, Technische Universiteit Eindhoven
Prof. dr. ir. G.S. Stelling, Technische Universiteit Delft

ISBN:1-58603-638-6

Keywords: operational water management, model predictive control, real time control, automation, centralized control, open water systems, irrigation, drainage, hydro-dynamic modeling.

Published and distributed by IOS Press under the imprint Delft University Press

Publisher & Distributor
IOS Press
Nieuwe Hemweg 6b
1013 BG Amsterdam
Netherlands
Fax: +31-20-687 0019
Email: info@iospress.nl

Distributor in the USA and Canada
IOS Press, Inc.
4502 Rachael Manor Drive
Fairfax, VA 22032
USA
fax: +1-703-323 3668
e-mail: sales@iospress.com

PRINTED IN THE NETHERLANDS

Summary

Human life depends on water daily, especially for drinking and food production. Also, human life needs to be protected against excess of water caused by heavy precipitation and floods. People have formed water management organizations to guarantee these necessities of life for communities. These organizations manage a water system within the community and manipulate the water flows in this system to fulfill the water related requirements. To do so, controllable structures, such as gates and pumps are used. The way these structures are controlled, depending on the requirements of the communities, is part of the research field of control on water systems, often referred to as operational water management.

In the research 'Model Predictive Control on Open Water Systems', the relatively new control methodology Model Predictive Control is configured for application of water quantity control on open water systems, especially on irrigation canals and large drainage systems. The methodology applies an internal model of the open water system, by which optimal control actions are calculated over a prediction horizon. As internal model, two simplified models are used, the Integrator Delay model and the Saint Venant model. Kalman filtering is applied to initialize the internal models. The optimization uses an objective function in which conflicting objectives can be weighed. In most of the cases, these conflicting objectives are keeping the water levels at different locations in the water system within a range around setpoint and executing this by using as little control effort or energy as possible. To tune the weight factors in the objective function, an estimate of the maximum allowed value of each variable in the objective function is used. The optimization takes the constraints of the control structures into account. Every control time step, the optimal control actions are calculated, while only the first set of control actions is actually executed. This results in a controlled water system that is constantly maintaining the objective in an optimal way, while taking predictions, such as expected irrigation demands or extreme storm events, and the constraints of the water system into account.

To show the potential of Model Predictive Control in controlling water systems, it is compared to the classical control methods Feedback Control and Feedforward Control. This comparison shows that Feedback Control has the lowest performance, as it first requires a deviation from setpoint to actually start the control actions. Adding Feedforward Control improves the performance. Many water systems are subject to constrained controllability of the structures. For example, pumps have a limited pump capacity and the flow through gates can be limited by the (sea)water level next to the structure. Model Predictive Control takes these constraints into account while calculating the optimal control actions. For that reason, Model Predictive Control outperforms Feedback Control and Feedforward Control in periods of extreme load. In other periods, the performance of Model Predictive Control is at least comparable to the performance of Feedback Control in combination with Feedforward Control. Another advantage of Model Predictive Control is the ability of dealing with conflicting objectives. In the objective function, relative weights are given to the different objectives, resulting in a well-balanced set of control actions for the total water system and potential

problems, such as an excess or lack of water, can be dispersed throughout the system as much as possible.
The optimization used in this research is suitable for linear internal models. As the water flows in canals and the structure flows are non-linear, a step-wise linearization is used in the optimization. By applying a number of iteration steps, the step-wise linearization approaches the non-linear solution with sufficient accuracy. In this way, all non-linear objects in a water system, even strongly non-linear structures such as pumps switching off and on, can be taken into account in the optimization.
An extension to the standard Model Predictive Controller is applied, by which uncertainties in predictions and models can be dealt with. Instead of optimizing one model, three parallel models are used in the internal model of the optimization. One model represents the average, most probable case. The other two models correspond to the best and the worst case. By multiplying the outcome of the three models by their probability of occurrence, the risk of high water levels is minimized, instead of minimizing high water level for just one of the possible cases. This stochastic configuration of Model Predictive Control is referred to as Multiple Model Predictive Control.
Model Predictive Control and its derived configurations are applied to accurate models of open water systems and on actual irrigation canals and drainage systems in real-time. The results show a clear improvement compared to the classical control methods. As the controller is set up in a generic way, it can easily be adapted to other water related fields, such as water quality control or water-power generation and on other types of water systems, such as reservoirs and sewer systems.
Finally, it is important that the solution to a control problem has to be as simple as the requirements on the controlled water system, the characteristics of the water system and the constraints of the structures allow it to be. In many cases, local Feedback Controllers have sufficient performance. In other cases, the application of the more complex Model Predictive Controller is unavoidable. A selection procedure for the most appropriate controller is part of this research.

Preface

For me, life is an ongoing quest for knowledge and wisdom. With this dissertation, I gained a lot of knowledge. I want to continue my life learning, especially gaining wisdom. From the people around me I have learned the following lessons:

Rob Brouwer	Irrigation technology, how to treat graduated students as part of the big water management family
Okko Bosgra	Control theory, the importance of real-time tests to proof the correct functioning of control methods
Sjoerd Dijkstra	Control theory, the way humor puts serious work into perspective
Guus Stelling	Fluid mechanics, numerical solutions of water flows, that there is life after setback, though it takes some beer-nights out
Bert Clemmens	That working hard and writing many scientific articles establishes your name and fame
Jan Schuurmans	Feedback control on open channels
Charles Burt	Irrigation practice
Wytze Schuurmans	Entrepreneurship
Betty Rothfusz	That being part of a work group feels better that working at home all the time
Bob Strand and Karla Strand	Hospitality, English language, appreciating a good Belgium beer
Lars van Elsakker, Carolien van Kampenhout, Dirk-Jan van Eck and Jan Bolders	Love for music, friendship
Jan-Willem Bakker	Humor, putting life into perspective, joy of making music together, what it means to be friends

Steven Weijs	That I am not always right with my technocratic ideas, importance of discussion, combining work and fun
Michèl van Leeuwen, Martijn van der Neut, Marcel Bruggers, Christian Huising, Willem Malda, Robbert Wagemaker, Paul Roeleveld, Bas van Rossum and Albert Goedbloed	The joy of working with intelligent students, combining work and fun
Patricia Woldring	That having a child together is the ultimate confirmation of true love, English language, Dutch language, interest for all cultures
Amanda van Overloop-Everaert and Cees van Overloop	The joy of working hard, that family is there to help you out under all circumstances
Birgit van Overloop	Justice, to stand up for the weaker in society
Martin van Overloop	How good it feels to be a father
Lola van Overloop and Wim van Overloop	To be a good and warm personality, that deceased people accompany you in your mind for the rest of your life, the joy of teaching.

Thank you all !!!

Table of Contents

1 Introduction

1.1 Managing water systems

Water plays an essential role in the life of every person. It is used for drinking, growing crops for food, sewerage, in processes to manufacture products, to generate energy and to carry ships over waterways. An additional aspect of water, especially in lowland areas, is that people have to be protected against extremely high waters caused by high tides, high river discharges and heavy precipitation.
As people live and work dispersed over large areas, the water needs to be distributed. This is illustrated by two examples; an irrigation system and a drainage system.

- Farmers have land in remote areas. They need fresh water to grow their crops. However, fresh water is often only available, from a distant source such as a reservoir or river. So canals are dug to convey the water from source to remote area, from supply to demand;
- When there is heavy precipitation, this water can flood land, causing damage and threatening lives. The excess water needs to be drained out of the area towards a river or sea. If the drainage capacity is restricted, it needs to be stored equitably in the distributed available storage in the area.

In both cases, water needs to be transported. To do so, people living together in an area have formed organizations that manage the water flows in the area. For the examples given, the organizations are respectively called irrigation districts and water boards. In order to manage the water flows, the organizations have transport canals and control structures at their disposal. At locations where control structures are present, flows can be influenced in order to manage the water flows in the canals. In general, this is done by operators who have built up experience in managing the water system. They work at a central location where they can have access to information about the state of the water system and predictions of future changes. For irrigation systems, these predictions can be the offtake schedules in which water orders are recorded. For drainage systems, a precipitation forecast is often used to predict future changes to the water system. If the present or the future state of the system is not in some desired state, the operators come up with adjustments to the structures to correct this. The operators at the central location communicate the required actions to operators that are responsible for operating the structures at the remote locations. These local operators change the settings of the structure, such as the position of a gate or the switching off and on of a pump. This method of managing a water system is referred to as central control. Other organizations operate the structures using only information of the remote location itself. This is referred to as local control. As central control uses information from multiple locations and can make

adjustments at multiple locations in the system, it is generally superior to local control in bringing the entire system in the desired state.
The operators usually describe the desired state of the water system in terms of target water levels (setpoints). The reason is that water levels are easily (visually) measurable and they do not change rapidly. If water levels in a supply system are kept to target, the supply to the off-taking users is generally assured. If water levels in a main drainage system are kept to target, risk of flooding is averted and drainage systems can safely evacuate their excess water into the main drains. Another indication that water levels are the most important control variables is that any deterministic and heuristic formula used in water management contains the variable water level. This is illustrated by three examples.

- If the water levels in the ditches of agricultural land are too low, there is no flow towards the ground water table and the crops will suffer. Yields will be reduced;
- Formulas describing the damage or number of casualties due to inundation use water level as an indicator;
- The available storage volume of temporary storage reservoirs can be easily computed from the water level in the reservoir.

Water system management can be formalized in a general structure diagram as given in Figure 1.1. The feedback controller corrects for measured deviations from set point, while the feedforward controller uses an estimate of the disturbance to counter weight the influence of the disturbance on the water level in the open water system. This block diagram holds for both central and local control, although local control generally only applies feedback control. Note that this research focuses on water quantity challenges, but that water quality specifications are often translated in extra specifications on the water quantity management (Hof & Schuurmans (2000)).

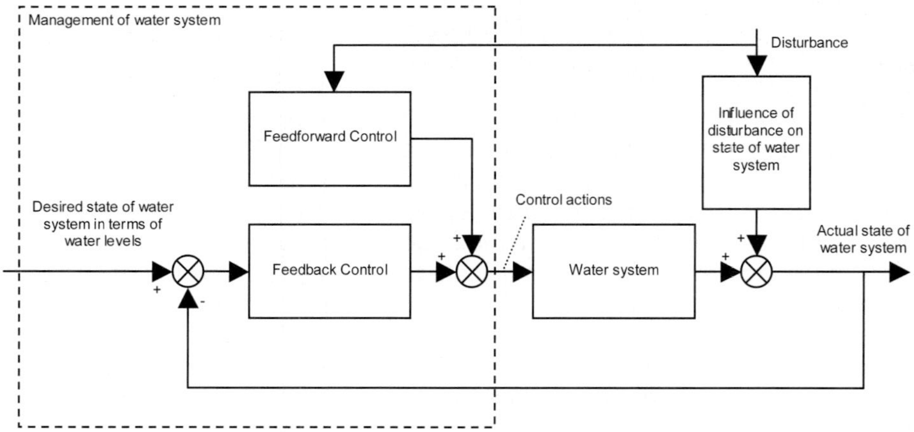

Figure 1.1 Structure diagram management of water systems

Although the main goal of operators is to maintain the water levels in the water system close to the target level, a secondary goal is to achieve this with minimal effort and cost. With manual adjustable gates the effort involves having a person to go out to the gate and execute the required adjustment to the gate, whereas with pumps the operators want these pumps to use as little energy as possible.

1.2 Automation in water management

The use of automatic control in managing open water systems has evolved slowly over the last decades. Two types of automation can be distinguished namely mechanical automation and electrical automation. The first type utilizes structures that are composed of floaters and levers attached to gates in such a way that a certain water management objective is achieved. These hydro-mechanical structures, such as Begemann-gate, Vlugter-gate and Neyrpic-AMIL gate (Brouwer (2004)), are able to maintain local upstream or downstream water levels close to a pre-defined target level. As this research focuses on complex, multi-variable water systems with changing dynamic behavior and changing water management objectives, these structures are not further investigated. Instead, the historic development of electrical automation is described. This development began with automatic water level sensors that register the water level at a certain sample rate. Next, these measurements became available at a central location through the use of communication lines. These lines are fast when direct cable or radio is used or slower if modems have to connect through telephone lines with the remote location. In some cases, additional information on predicted disturbances also became available. At the same time these communication lines were used for sending the required adjustments to the structures. The structures themselves have been increasingly automated with electric motors and automatic switching off and on of pumps. Organizations that have evolved to this level of automation posses a management system that is capable of implementing central control by using a central computer. For many plausible reasons most of the organizations have not evolved this far. In third world countries for example, labor is inexpensive, so manually adjusting structures is not at all costly (Burt (1999)). The main reason though, is that the central system in both ways completely depends on communication lines. From past experiences these lines have often proven to fail, causing the central control to fail completely. For this reason, many organizations manage the water flows by way of local control. Here, the full automation is configured as follows: automatic water level sensors, cable lines to a Programmable Logic Controller in which the adjustments to the structures are computed and cable lines back to the motorized structures (Burt & Piao (2002)). The choice of the configuration of the management system and of the appropriate control method to function adequately depends completely on the specifications of the controlled water system. These specifications include the following demands: the range of regular operating points; an indication of how much the water levels and flows may fluctuate around setpoint within these operating points; how long water levels and flows may be off target; and how often structures may be adjusted. Even though the operators are often vague about what the exact specifications are, after some discussion the specifications are in general stipulated as keeping the water levels as close to target level as possible with a minimum and maximum allowable water level as range around the target level and as few adjustments to the structures as possible with a minimum time the structures may not be changed. In case pumps are used, an extra requirement is to use as little energy as possible.

1.3 Control of water systems

Management of open water systems can be formalized in a set of logical and mathematical rules within a controller. In the past, the different types of controllers have been categorized in their water management related characteristics (Brouwer (2001)). One way of categorization is:

- Flow control. A certain flow rate is imposed at a structure by changing the structure settings, such as gate width or gate opening;
- Volume control. The volume in a canal reach is kept as close to a target volume as possible (Seatzu & Usai (2002)). This type of control can have benefits when the management of the canal reaches is subject to frequent shut downs in which the canal transits from steady flow to zero flow. As the volume can not be measured directly, more than one water level in the reach is measured, for example at the upstream and downstream end of the reach. From these water levels, the volume of water in the reach can be estimated by some weighing formula. This control method does not differ fundamentally from the next type of control;
- Water level control. The water levels in canal reaches are kept as close to a target water level as possible. This type of control is researched in this dissertation.

Another way of categorization is based on the location of the water level that has to be kept at target level relative to the control structure:

- Downstream control. The water level downstream of the control structure is kept as close to target level as possible. By applying this method, shortage of water in the downstream canal reach is replenished by extra inflow through the upstream control structure. This property makes upstream control highly suitable for control of irrigation systems. Note that the controlled location can be chosen further downstream in the canal reach. Especially steep canal reaches with embankments parallel to the bed slope, require their controlled point at the downstream side of the canal reach in order to avoid overtopping of these embankments. This type of downstream control is referred to as remote downstream control;
- Upstream control. The water level upstream of the control structure is kept as close to target level as possible. By applying this method, abundant water in the upstream canal reach causes the downstream

control structure to discharge extra water. This property makes upstream control highly suitable for control of drainage systems;

Various controllers have been implemented in practice. In some cases the entire control loop of measuring, computing the control actions and adjusting the structures is fully automated. In other cases, the controller is automatically supplied with measurements and this controller advices the operator. The operator can act according to this advice or can decide to ignore it and use his own judgment. This control system is referred to as Decision Support System (DSS). Many controllers have been designed for water systems, but have not yet been implemented in practice. They are tested though, on accurate hydro-dynamic models of the water system to prove their applicability. The various controllers for water systems can be categorized in the following general methods (Ruiz et al. (1995), Malaterre & Baume (1998a), Malaterre et al. (1998b)). Instead of the categorization based on the water management characteristics as presented above, the next classification is based on general control theory:

- Feedback control. Feedback controllers measure the water level, compare this level to the target level and compute the change in structure setting as a function of the deviation. Often this is a Proportional Integral controller (PI-controller) in which the change in structure setting is computed from a proportional gain factor and an integral gain factor multiplied by the change in error and the error itself, respectively. The values for the proportional and integral gain factor, found in a tuning procedure, determine the behavior of the controlled water system. If the gain factors are tuned well, the controlled system will be robustly stable, reasonably fast and without severe fluctuations in structure setting. The feedback controller constantly corrects the difference between measured water level and target level in a repetitive loop. For that reason, this control method is generally referred to as closed loop control. The deviation between water level and setpoint results from disturbances that influence the water level, such as offtake flows or storm events. In this way, the feedback controller functions as disturbance rejection (Vandevegte (1990), Schuurmans, J. & Liem (1995b), Schuurmans, J. (1997), Schuurmans, J. et al. (1999b));
- Feedforward control. Feedforward controllers use measurements or predictions of a disturbance and an inverse model of the effect this disturbance has on the water level, to compute the required adjustments to the structures. The feedforward control action aims to precisely cancel this effect which would ideally result in a zero water deviation from target level (Dhondia et al. (2000), Bautista et al. (2003), Wahlin & Bautista (2003)). This control method is generally referred to as open loop control. As the inverse model can never perfectly represent the inverse of the effect the disturbance has on the actual water level, measurements and predictions are often inaccurate and the dynamic behavior of the actual water system changes over time, this deviation will never be zero. A combination of feedback and feedforward is often used to have the feedback control action compensate for the imperfection of the feedforward control action (Schuurmans, W. et al.

(1999), Overloop et al. (2001), Mareels et al. (2003), Overloop (2003a), Roos (2003)), Steenis et al. (2003), Huising (2004), Montazar & Overloop (2005), Overloop (2005a));

- Optimal control. The most common optimal controllers in water systems are based on the Linear Quadratic Regulator theory. These optimal controllers minimize an objective function by using a numerical optimization algorithm. In the objective function the square of the deviation between water level and target level is weighted against the square of the change in structure setting. The square sign gives an equal penalty for both positive and negative deviations and structure adjustments. The relative weighing between water level deviation and structure adjustment is found in a tuning procedure. By changing the weight factors, higher penalties can be set on the water level deviation or the structure adjustments, resulting in faster control (smaller deviations) or smoother operation of the structures respectively (Kwakernaak & Sivan (1972), Reddy (1990), Malaterre (1995) , Malaterre & Rodellar (1997a), Clemmens & Schuurmans, J. (2004), Clemmens et al. (2005));
- Heuristic control. Opposed to the first three deterministic control methods, a group of control methods can be distinguished that is not based on physical laws, but uses a more heuristic approach. Examples of these methods are control based on rules-of-thumb, neural networks control, fuzzy logic control and genetic algorithm control. Control based on rules-of-thumb are common for water systems that can be controlled in a straightforward, standardized manner and are not subject to control objective changes over time. Neural network control can be used if a large amount of measurements of water levels and control actions is available and the water system is too complex to model with physical formulas. Fuzzy logic can be relevant when the behavior of multiple operators working on the same control task needs to be reproduced. Genetic algorithms can find an optimal solution faster than numerical deterministic optimization algorithms. For large optimization problems though, this solution is often a local optimum. A drawback of all these methods is that the dynamic behavior of water systems is seen as a black box. Especially on this behavior, extensive research has been done over the last century and accurate formalizations of this behavior are available. These heuristic methods are not applied to a large extent in controlling open water systems.

Model Predictive Control presented in this research, has elements of the first three control methods namely feedback control on measured water levels, feedforward by using measured and predicted disturbance and optimal control to allow for high performance control of large water systems with interconnected canal reaches.

1.4 Problem statement

In many cases, the control methods previously described can satisfy the specifications that are given for a controlled open water system. However, in other cases there is a limiting factor that makes it impossible for these control methods to function in a satisfactory manner. This factor is the limited capacity of the structures and transport canals that are used. These limited capacities are referred to as constraints on the system. The limited capacity can also become relevant if the specifications of the controlled system become more stringent over time. This higher requirement is unavoidable, as history has shown that the socio-economic demands of the society in which the water system functions increase over time. The influence that constraints have on the controlled behavior can be illustrated by two examples; a controlled irrigation system and a controlled drainage system:

- In the past, farmers in dry areas depended on the water that was available to them irrespective of what time of the day that was. This type of water supply led to low efficiency and low performance of the distribution system. Nowadays though, farmers have gained more political power and want their water on demand. If all farmers start to irrigate at 7 o'clock in the morning the capacity of the canals and the structures might not be high enough to accommodate the large step in flow change. The way the operators deal with this problem is to store more water in the canals before the offtake change takes place.
- Drainage systems in lowland areas sometimes have to deal with extreme storm events. The operators have to keep the water levels below a certain maximum level. To achieve this, they have pumping stations with limited capacity at their disposal. If the runoff caused by extreme storms is higher than this capacity plus the available storage between target level and maximum allowable water level, this will result in inundation and consequently in damage. As operators have predictions of the storm event available, they avoid this problem by temporarily lowering the water level in the drainage canals. They prematurely will start pumping out water a couple of hours before the storm event actually takes place.

In both examples, the operators use the effect a prediction has on their control target and the fact that the controllability is limited by the constraints on the structures they operate. It is clear that for complex water systems with interacting subsystems, water management including feedback, feedforward, weighing of small water level deviations against minimal structure adjustments and constraints on structures becomes a difficult, if not impossible task. Here, control theory comes into play to support the water manager in a formalized and systematic way.
To effectively control water systems that are characterized by optimization problems and constrained structure capacities, more advanced control methods are required. From other engineering fields, especially from control of chemical plants, a certain control methodology has gained more and more popularity over the last few decades. This control methodology is most commonly referred to as Model Predictive Control (MPC). This methodology combines feedback control on the measured water levels and feedforward control on the predicted disturbances in a repetitive optimization procedure that also takes the constraints on the structures into account. In Figure 1.2 the structure diagram of a water system

controlled by Model Predictive Control is presented. The same inputs and output as the water management structure diagram (Figure 1.1) are used with addition of an objective function and the constraints.

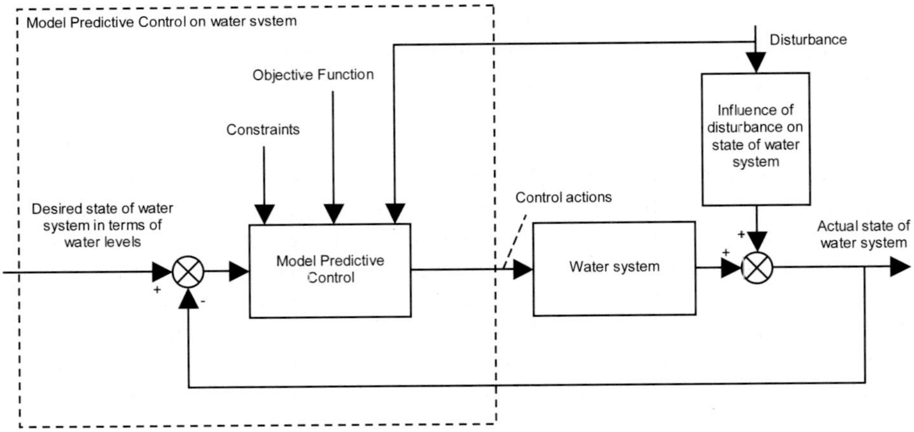

Figure 1.2 Structure diagram Model Predictive Control on water systems

Tests with Model Predictive Control methods on simplified models of water systems have shown promising results, although they do not solve all practical problems and lack a standardized formalization for various types of water systems (Zagona (1992), Ellerbeck (1995), Lobbrecht (1997), Malaterre & Rodellar (1997b), Eklund & Tufvesson (2001), Gomez et al. (2002), Wahlin (2002), van Leeuwen (2003), Wahlin (2004), Glanzman et al. (2005)).
Now, the research goals of this dissertation can be formulated as follows:

- To analyze the challenges of water quantity management of open water systems;
- To configure a standardized Model Predictive Control formalization for various types of open water systems (both irrigation systems and drainage systems) that incorporates those water management challenges;
- To analyze the limitations of present control methods and to show that Model Predictive Control does not have this limitation. To demonstrate the increased performance, MPC is applied to various types of open water systems subject to realistic water management requirements under realistic circumstances.

1.5 Outline of dissertation

To analyze the possibilities of Model Predictive Control on open water systems, the components of controlled open water system need to be analyzed in detail.

These components are the sub-systems of the open water system that are part of the closed control loop. Additionally, previous research has shown that these components have a great impact on the stability and performance of the controlled water system. In Chapter 2, the background on the dynamic behavior of open water systems and its components is presented. The dynamics are captured in models to be able to work with model based controllers such as MPC. Chapter 3 describes the setup of a Model Predictive Controller that deals with the specific water system related challenges. In Chapter 4, the use of MPC in a sequential loop is explained. With this configuration, non-linearities and off/on structure settings in the control problem can be assessed. Chapter 5 presents a derived configuration of MPC that utilizes multiple models as internal model. This configuration allows for a stochastic approach to the control problem in which uncertain models are used. All these chapters capture the theory that has been organized to come to a formalized and standardized Model Predictive Controller that can deal with the challenges of managing open water systems in a systematic way. Next, MPC and the derived configurations are applied to actual open water systems and accurate models of open water systems to show the proper functioning and the benefits of this control methodology. Chapter 6 contains these applications on various types of open water system. In Chapter 7, the conclusions are presented, as are suggestions for future work in this field.

2 Open water systems

2.1 Formalizing open water systems

The research goal of this dissertation is to improve the management of open water systems by applying model based control. To do this in an efficient way, the models of these water systems need to be set up in such a way that they contain the dynamics that are relevant for this management. In most cases these dynamics are the water levels in canals at various locations and the water flows that influence these water levels. By using structures to manipulate the flows, a controller can achieve the management objective. This is to keep the water levels as close to setpoint as possible, even though boundary conditions such as tidal water level boundaries or varying in- and outflows disturb the water system. By modeling all these parts of the water system (canal reaches, structures, disturbances, controller) a model based controller can predict the future water levels and flows that are the result of the disturbances and the control actions (Bosgra (2003)). In the next chapter, the model based controller referred to as Model Predictive Control is described. This present chapter formalizes all sub-models that can be put together to form the entire water system model.
Two general types of water systems are considered in the Model Predictive Controller method that is applied in this research namely irrigation and drainage systems. In Figure 2.1 an imaginary but realistic drainage system is shown. It contains sub-systems that are described in detail in the following paragraphs. The system consists of two flat canal reaches. The first reach R1 has rainfall-runoff inflow at the upstream side. The water flows to the downstream side, where a pump station switches on if a certain level is exceeded by the water level. The pump lifts the water to the next reach R2. In this reach the water flows towards the undershot gate that drains the water out of the system. The boundary at the outside of the system is a sea water level tide. During low tide the water level in the canal reach is not influenced by the outside water level i.e. the gate is free flowing. In case the sea water level becomes higher, the undershot gate submerges and the flow through the structure decreases. Once the sea water level is higher than the canal water level, the gate is closed to avoid intrusion of saline water into the canal.

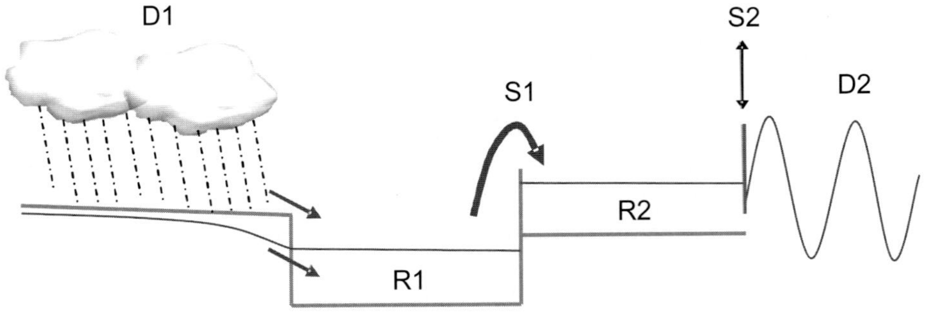

Figure 2.1 Drainage system with sub-systems

In Figure 2.2 an imaginary, but realistic irrigation system is presented. The irrigation canal consists of one steep canal reach (R3) and two flat canal reaches (R4, R5). In the first reach the water flow is super-critical up until the part that is in back water. The structures are respectively, a free flowing undershot gate as head gate, another free flowing undershot gate, a submerged undershot gate and a free flowing overshot gate as spillway at the end of the canal. At the downstream side of each reach an outlet is located. These offtakes are usually undershot gates towards secondary canals or small pumps pumping the water directly upon the fields. The flow that the offtakes will receive is often recorded in an offtake schedule.

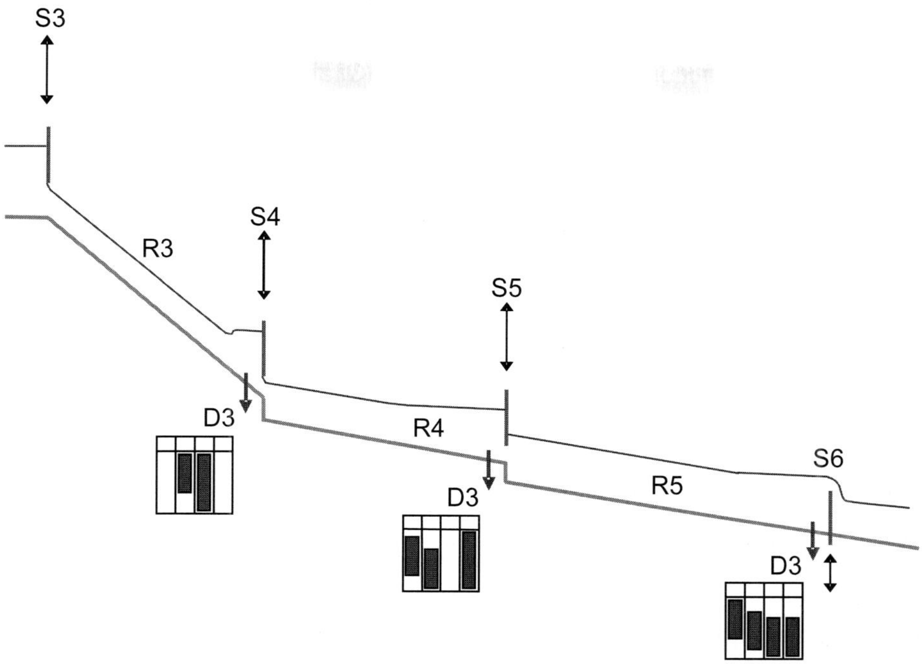

Figure 2.2 Irrigation system with sub-systems

2.2 Canal reaches

This paragraph describes the relevant dynamics of canal reaches. Relevant dynamics of the reaches are the water movements that make the water levels drift away from setpoint, and for which a controller effectively has to correct. Any model based controller must be designed taking all relevant dynamics into account. In this paragraph two types of actual canal reaches are demonstrated in a detailed hydro-dynamic model (Sobek (2000)). One is a flat and deep canal reach, while the other is steep and shallow. Each demonstration reach has its own specific dynamics that has to be taken into account when designing a controller.

2.2.1 Modeling of canal reaches

The water in canal reaches flows, driven by gravitational forces, along the meandering of the reach from the upstream to the downstream side. The velocity and the amount of the one-directional flow at each location along the stretch depend on the dimensions of the canal reach at these locations. The dimensions that influence the flow are the cross sectional area, the steepness and the roughness of the bed. A general accepted way in literature of describing the water levels and water flows in shallow canal reaches is by using the De Saint Venant equations (Chow (1959), Cunge et al. (1980), Stelling & Booij (1994)). These equations are partial differential equations that consist of a mass balance and a momentum balance. The momentum balance is a summation of the descriptions for the inertia (1), advection (2), gravitational force (3) and friction force (4).

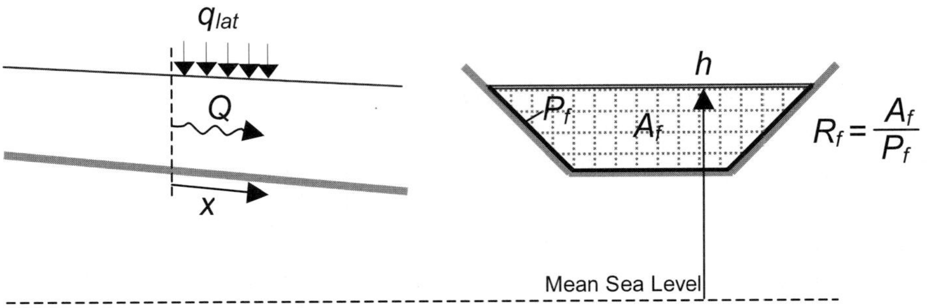

Figure 2.3 Canal reach schematisation

$$\frac{\partial Q}{\partial x} + \frac{\partial A_f}{\partial t} = q_{lat} \qquad \text{Formula 2.1}$$

$$\underbrace{\frac{\partial Q}{\partial t}}_{(1)} + \underbrace{\frac{\partial}{\partial x}\left(\frac{Q^2}{A_f}\right)}_{(2)} + \underbrace{g \cdot A_f \frac{\partial h}{\partial x}}_{(3)} + \underbrace{\frac{g \cdot Q|Q|}{C^2 \cdot R_f \cdot A_f}}_{(4)} = 0 \qquad \text{Formula 2.2}$$

where Q represents the flow (m^3/s), t the time (s), x the distance (m), A_f the wetted area of the flow (m^2), q_{lat} the lateral inflow per unit length (m^3/s/m), g the gravitational acceleration (=9.81 m/s^2), h the water level (mMSL), C the Chézy friction coefficient (m$^{1/2}$/s) and R_f the hydraulic radius (m). R_f is calculated by A_f over P_f, where P_f represents the wetted perimeter (m).
To use these formulas in a numerical model of a canal reach, the partial differential equations are discretized in time (Δt) and space (Δx). In case these discretized formulas are simulated, the model results in time series solutions of water levels and flows at discrete locations along the reach. Also, the time series are discrete solutions in time. Often the De Saint Venant equations are discretized with the Preissmann scheme (Cunge et al. (1980)). A more numerically robust method of discretizing the equations is by using a staggered grid with wind-up implementation (Stelling & Duinmeyer (2003)). An example of the higher robustness compared to the Preissmann scheme is that it can deal with super-critical flow (R1 in Figure 2.1). This robust discretization is used in this research.

Both the steep and the flat reach are used to show the solution in time at all locations along the canal reach. The dimensions of the steep canal are given in Table 2.1. In Figure 2.4 the steady state of low flow and high flow are presented as calculated with a hydro-dynamic modeling package (Sobek (2000)).

Table 2.1 Dimensions of steep canal reach

Parameter	Value	Unit
Length	500	m
Bed slope	4e-3	-
Bottom width	1.22	m
Side slope	1.5	-
Friction	0.02	Manning $s/m^{1/3}$
Water depth setpoint	0.85	m
Flow at start (low flow)	0.36	m^3/s
Flow at end (high flow)	1.44	m^3/s

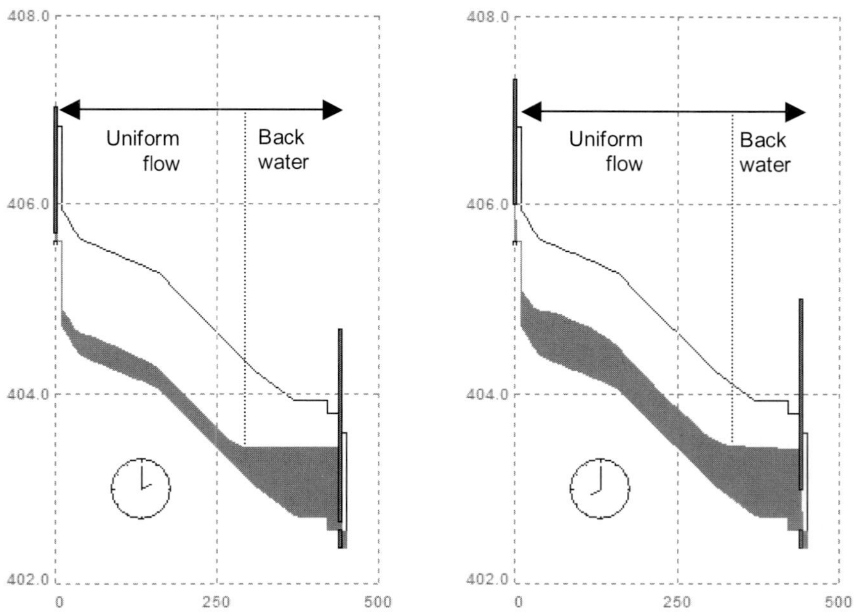

Figure 2.4 Numerical solution of the De Saint Venant equations in a steep canal reach at a lower and higher flow

The steady state solutions at the initial and the final situation when the water has settled again, consist of two parts namely the uniform flow part with normal depth and the back water part with increasing depths and a more or less horizontal water surface. As the flow through the canal reach is higher at the end of the

simulation, the normal depth is greater and the point of intersection between the uniform flow part and the back water part has moved further downstream.
The dimensions of the flat canal are given in Table 2.2. In Figure 2.5 the solution in time of the flat canal reach is shown. The reach is disturbed by a simultaneously stepwise increase in flow at both the upstream side and the downstream side. The pictures represent 3.5 minute time intervals, which coincides with the repetitive high and low water level at the downstream side of the reach.

Table 2.2 Dimensions of flat canal reach

Parameter	Value	Unit
Length	1100	m
Bed slope	1e-3	-
Bottom width	10	m
Side slope	1.5	-
Friction	45	Chézy $m^{1/2}/s$
Water depth setpoint	4.27	m
Flow at start	50	m^3/s
Flow at end	75	m^3/s

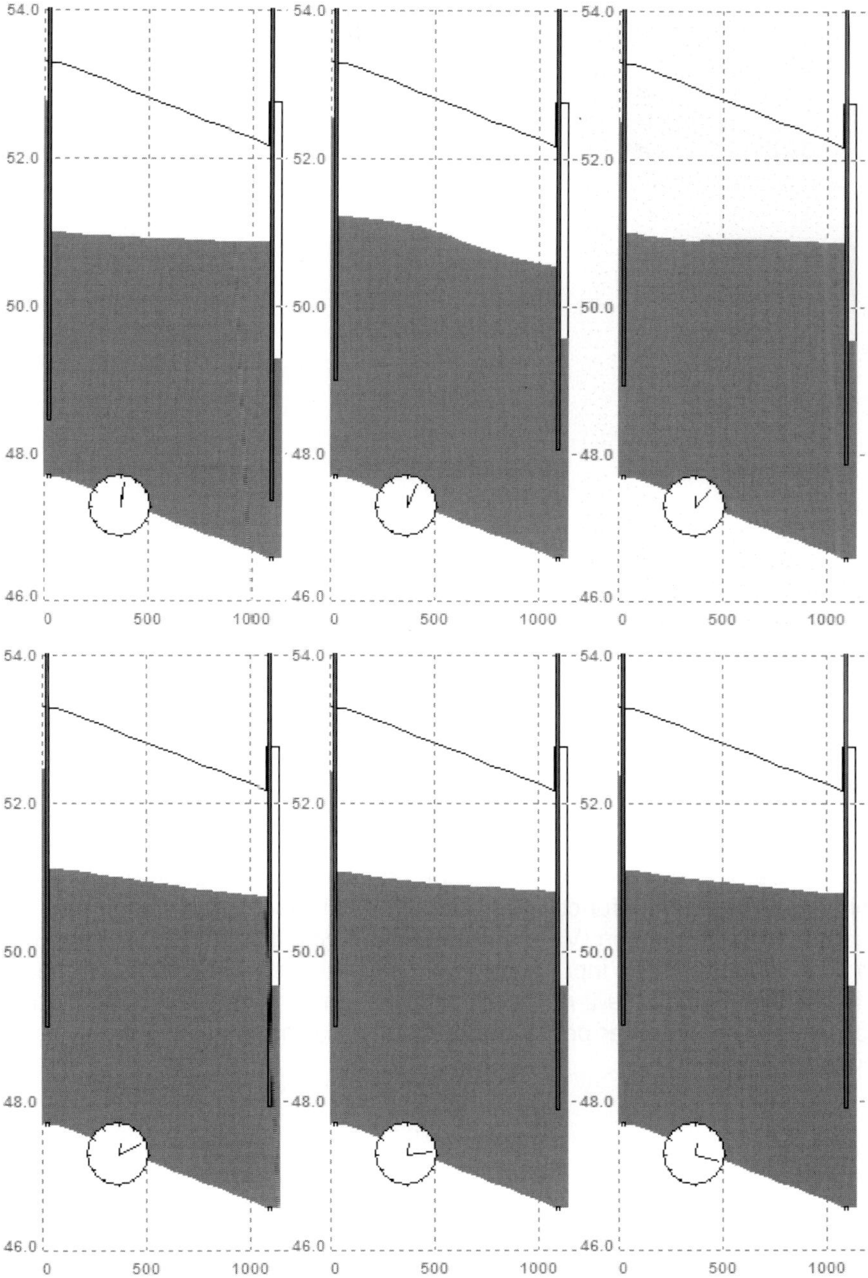

Figure 2.5 Numerical solution of the De Saint Venant equations in a flat canal reach

The simulation starts at 1 minute from steady state flow. At 18.5 minutes the flow has settled again. The intermediate solutions show a resonance wave with a frequency ω of 1.5e-2 rad/s. This is the reflecting wave that moves up and down the canal reach a number of times before it settles. The frequency ω can be estimated from the length, velocity and celerity as calculated in Formulas 2.3 to 2.5 (Chow (1959)).

$$c = \sqrt{g \cdot d}$$ Formula 2.3

$$v = \frac{Q}{A}$$ Formula 2.4

$$\omega = \frac{2\pi}{\frac{L}{c+v} + \frac{L}{c-v}}$$ Formula 2.5

where *c* represents the celerity (m/s), *d* the water depth (m), *v* the velocity (m/s), ω the frequency (rad/s) and *L* the length (m). With the values given in Table 2.1 this results in a frequency ω of 1.80e-2 rad/s. The difference with the simulated frequency is approximately 20%. The difference is caused by the non-linearities of the water movements that are not taken into account in the Formulas 2.3 to 2.5. For example, the depth is smaller than the setpoint depth at locations more upstream of the controlled water level. This also causes a higher velocity of the water flow at those locations.

When the non-linear discrete De Saint Venant equations are linearized around a certain operating points (Bosgra (2003)) and analyzed in the frequency domain (Vandevegte (1990)), the resonance wave can be seen as the first peak M_{p1} in the magnitude of the transfer function as presented in Figure 2.6. This figure represents the transfer function (Vandevegte (1990)) with the inflow at the upstream side of the reach as input and the water level at the downstream side as output. The frequency is 1.28e-2 rad/s, which is less than 15% difference with the simulated frequency. The other peaks are the higher harmonics of this basic frequency.

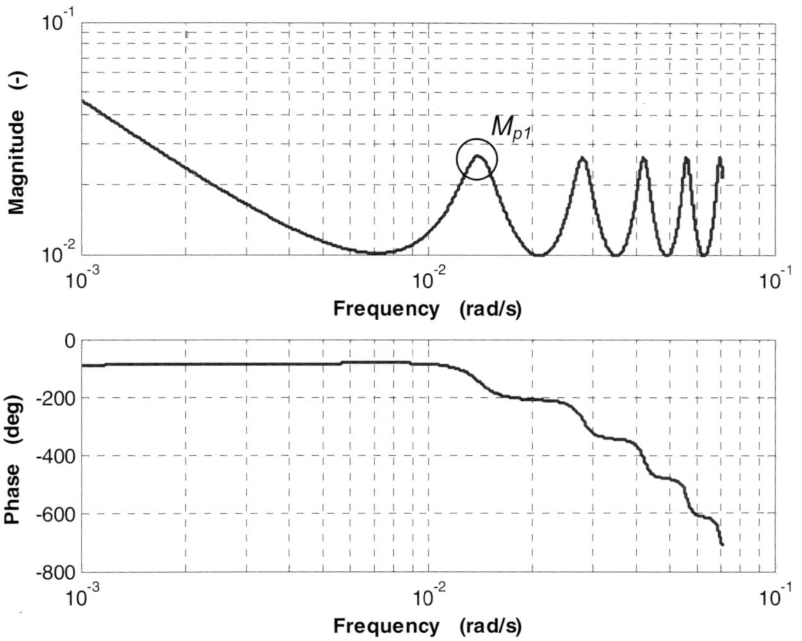

Figure 2.6 Bode diagram of linearized De Saint Venant equations

The resonance of the basic frequency is problematic for feedback control when the water level downstream in the canal reach is controlled by the upstream structure. In that case, the measurement is in counter phase with the control action, so the phase lag is -180°. If the open loop gain of controller and canal reach in series is larger than 1, the closed loop system becomes unstable according to the Nyquist stability criterion (Vandevegte (1990)). This means that, in general, canal reaches that have high resonance peaks are harder to control than comparable canal reaches with a more damped behavior. To be able to estimate the sensitivity of canal reaches for resonance waves, the computation of the first resonance peak and resonance frequency of the flat demonstration reach is repeated a number of times with varying parameters. In each trial, one of the parameters of the reach dimensions is changed with 10%, first 10% higher, then 10% lower. The results are summarized in Figure 2.7.

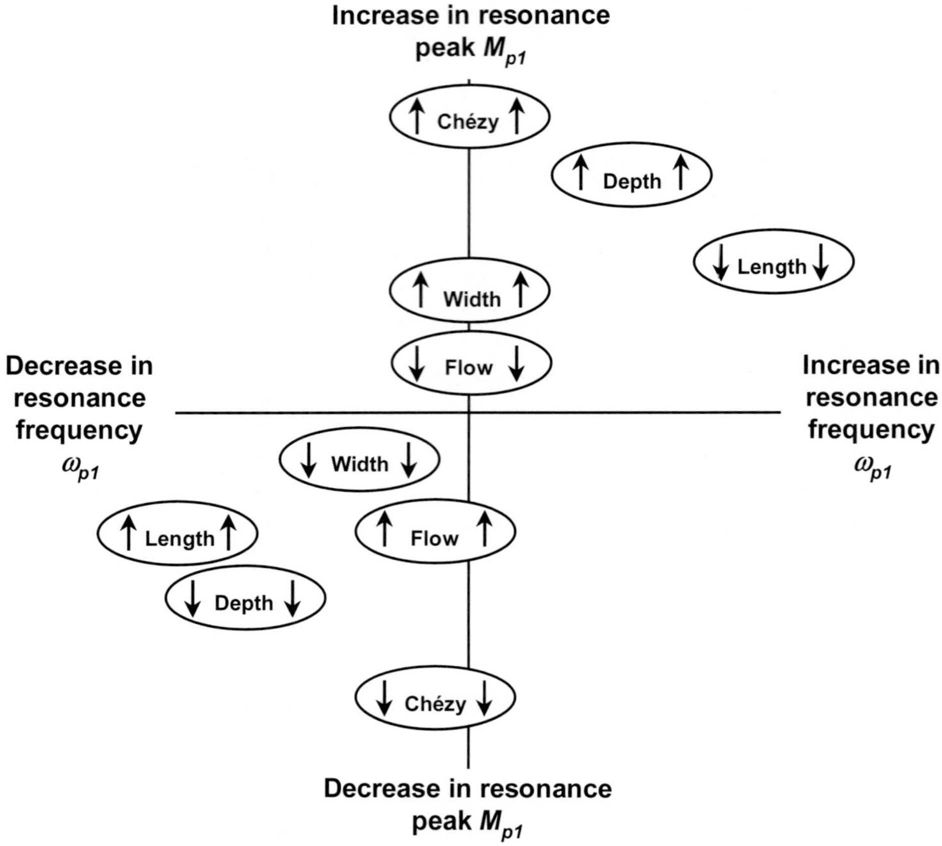

Figure 2.7 Influence of changes in parameter values of reach dimension on basic frequency

Upward arrows mean that the value of the specific parameter is increased with 10%. Downward arrows mean a decrease with 10%. The non-linearity of the water movements can be seen from the fact that the 10% increase and 10% decrease do not result in the same increase or decrease of the resonance peak and resonance frequency. Additionally, the resonance frequency ω_{p1} is mainly influenced by the length and the depth. This conclusion is in agreement with the Formulas 2.3 to 2.5, as the celerity c is mainly determined by the water depth d. The celerity and length L are the main parameters that determine the resonance frequency. A third conclusion is that the resonance peak M_{p1} is mainly determined by the friction force part in the momentum balance (part 4 in Formula 2.2). This part is determined by the flow Q, Chézy friction coefficient C, wetted area A_f and hydraulic radius R_f. The parameters depth d and width W that are varied in the sensitivity analysis are the main parameters in the calculation of the wetted area and hydraulic radius. Note that a higher Chézy friction coefficient results in a lower friction. The friction force can be seen as a damping component that works

in opposite direction of the flow. A low friction force results in less damping and, consequently, in high resonance peaks, while a high friction force results in low resonance peaks. These results show, that short, wide, flat (which usually means deep along the entire reach), smooth canal reaches (e.g. R1 in Figure 2.1 and R4 in Figure 2.2) at low flow are in general more sensitive to resonance waves than long, steep (which usually means shallow along the major part of the reach), rough canal reaches (e.g. R3 in Figure 2.2) at high flow.
As water systems convey water over long distances, they are characterized by significant delay times. The delay time τ of a controlled canal reach is the time between adjustment of the control structure and the change in controlled water level. This characteristic can have a large impact on the stability of the closed loop control of the canal reach. The higher the delay time, the more phase lag is added to the open loop behavior. This decreases the phase margin and makes the higher frequencies move towards the instability point of magnitude 1 and phase lag -180° (counter phase). With a flow step change ΔQ at the upstream side, the delay time from the upstream to the downstream side can be estimated. Figure 2.8 shows the visual determination of the delay time from the point of intersection between the rise line of the downstream water level and the horizontal setpoint line. The test is done with the steep demonstration canal reach.

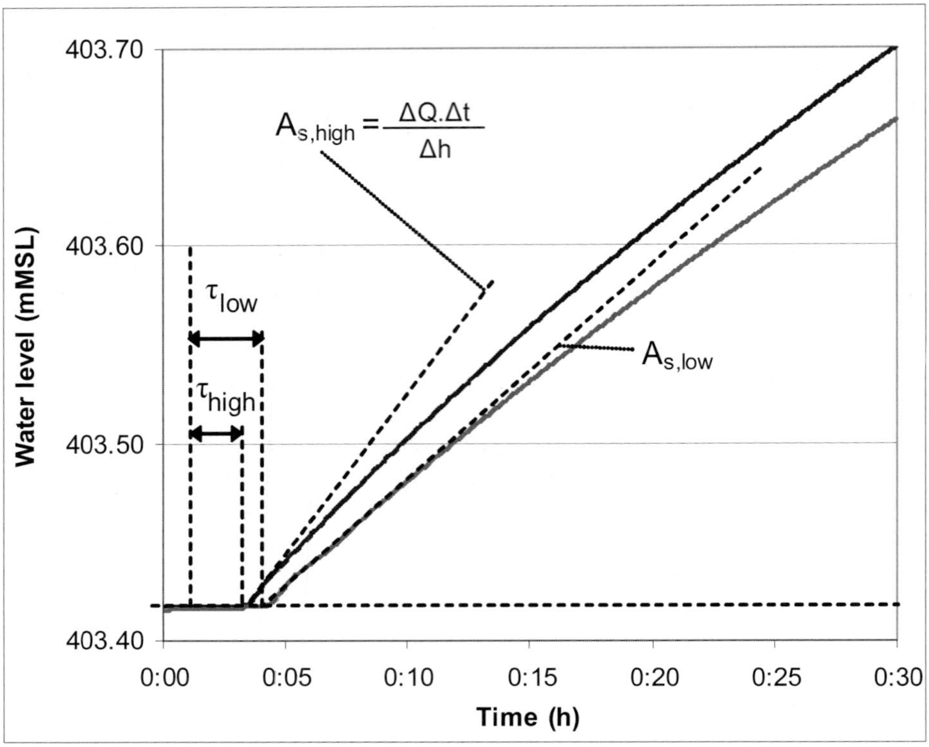

Figure 2.8 Estimate of delay time and storage area of canal reach

The delay times of the canal reaches determine the minimum length of the time horizon over which the model in the model based controller needs to be used. The effect of a control action upstream in a canal on the water level downstream in the canal needs to be taken into consideration. Consequently, for long canals consisting of canal reaches in series, the summed delay times of all reaches is usually taken as the prediction time horizon.

2.2.2 Simplified models of canal reaches

The discretized De Saint Venant equations can simulate the water levels and water flows in an open water system very accurately when the discretization is done with a small grid size Δx and a small time step Δt and when the resulting model is calibrated from actual measurements. An example of this is shown in Figure 2.8. In general, the calibration is performed by varying the parameter that is most difficult to estimate in advance videlicet the bed friction.
The detailed discretization model results in long computation times, which makes it difficult to use these models in real-time applications. Instead, simplified models are used. In the Model Predictive Control application as presented in Paragraph

6.3, a simplified model is applied, in which a discretization of the De Saint Venant equations is used with a large grid size and a large time step. This model is still accurate enough to capture the basic dynamics, such as the delay time and the basic frequency. The advantage of this simplified model is that it is valid at all operating points from low to high flow.
In the various applications of the Model Predictive Controller as described in Paragraph 6.1, 6.2, 6.4 and 6.5 the Integrator Delay model is used (Schuurmans, J. & Ellerbeck (1995a), Schuurmans, J. (1997), Schuurmans, J. et al. (1999b)). The discrete transfer function *H(z)* (Vandevegte (1990)) from upstream inflow Q_{in} and downstream outflow Q_{in} to downstream water level h as used in this model, consists of a delay time in series with an integrator part.

$$H(z)=\frac{h(z)}{Q_{in}(z)}+\frac{h(z)}{Q_{out}(z)}=\frac{z^{-k_d}}{A_s\cdot(z-1)}+\frac{1}{A_s\cdot(z-1)} \qquad \text{Formula 2.6}$$

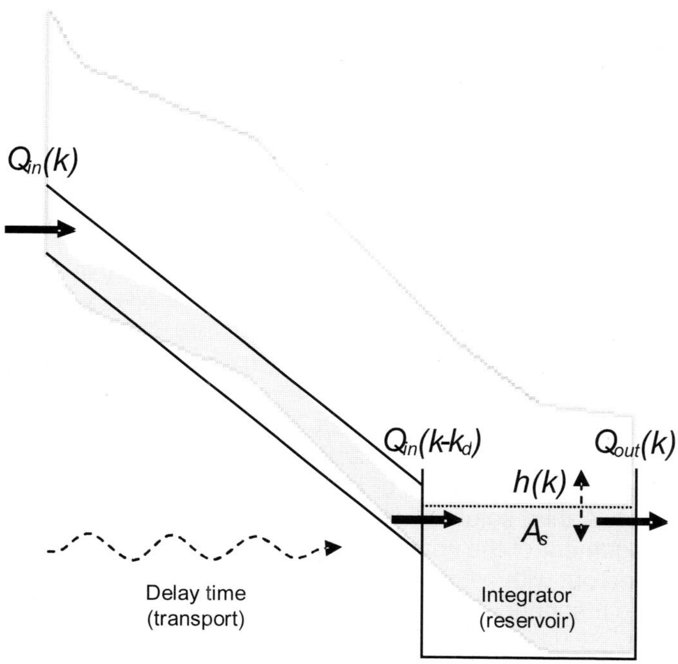

Figure 2.9 Integrator Delay model of canal reach

This model captures the delay time steps k_d and the storage area A_s by which the back water area moves up and down. z is the discrete operator for moving one step into the future. Figure 2.9 gives the physical interpretation of these two characteristics of a canal reach. The strength of the model is that it is very compact and therefore fast. The disadvantage is that the parameters delay time and storage area are only valid at one working point. The values of these parameters change when the flow changes. Table 2.3 gives an example of the irrigation canal with 8 reaches as tested in Paragraph 6.5. Reach 3 of this canal is the steep test reach as demonstrated in this paragraph. The values of τ and A_s at low, average and high flow are given in the table. The largest difference for the delay time is 63%, while the largest difference for the storage area is 52%. An additional disadvantage of the Integrator Delay model is that it does not contain the resonance frequency.

Table 2.3 Reach characteristics in various operating points

Reach	Low flow		Average flow		High flow	
	τ (s)	A_s (m^2)	τ (s)	A_s (m^2)	τ (s)	A_s (m^2)
1	0	397	0	379	0	343
2	534	653	360	600	288	450
3	120	503	90	493	78	240
4	162	1530	72	1621	60	1506
5	1152	171	828	240	702	248
6	792	1614	648	1385	540	878
7	540	2000	576	1385	504	1286
8	1008	1241	954	1319	720	1263

It is important to realize that, if the model based controller does not contain all relevant dynamics of the actual water system, the controlled water system in closed loop can become unstable. There are various ways to deal with this problem. One solution is to work with a model based controller that does not have a fixed internal model, but is time-variant with the operating points. The sequential configuration of Model Predictive Control as described in Chapter 4, allows for that. Another way of dealing with the changing operating points is to use multiple models in parallel in the controller. These local models are valid in the different operating points. Overloop et al. (2005b) give a practical application of this method. Also, the multiple model configuration of MPC as described in Chapter 5 can be used for this purpose. A third solution is to filter out the frequencies that are not of interest for management of the canal reach. In Schuurmans, J. (1997) the basic frequency is filtered out with a first order low-pass filter that is applied to reaches that are sensitive to resonance waves.

2.2.3 System identification of simplified models

There are various ways to identify the model characteristics. Figure 2.7 already showed a simple example of how to find the delay time visually. The storage area can be visually estimated by taking the tangent of the water level rise line. An automated way of finding these characteristics is to use an auto-regression estimate technique on an input and output signal of a canal reach. The input signal is a Pseudo Random Binary Signal (Ljung (1987), Silvis (1997), Silvis et al. (1998), Weyer (2001)) as switching flow at the upstream structure, while the

output signal is the resulting fluctuating water level at the downstream side of the canal reach. Malda (2005) has done this on the actual canal as tested in Paragraph 6.5. The goal of this identification is to capture the low-frequency response being the values of the storage area A_s and the delay time τ. For identification of the higher frequencies such as the basic frequencies and higher harmonic frequencies, more detailed tests are required. These tests must be preceded by an analysis of the frequency band that is to be identified.
In addition to the parameter estimate of the actual canal reaches, Malda (2005) also compared these results with the same test run on the detailed hydro-dynamic model of the canal reaches. The result of this comparison performed on the steep demonstration canal reach 3 is given in Figure 2.10.

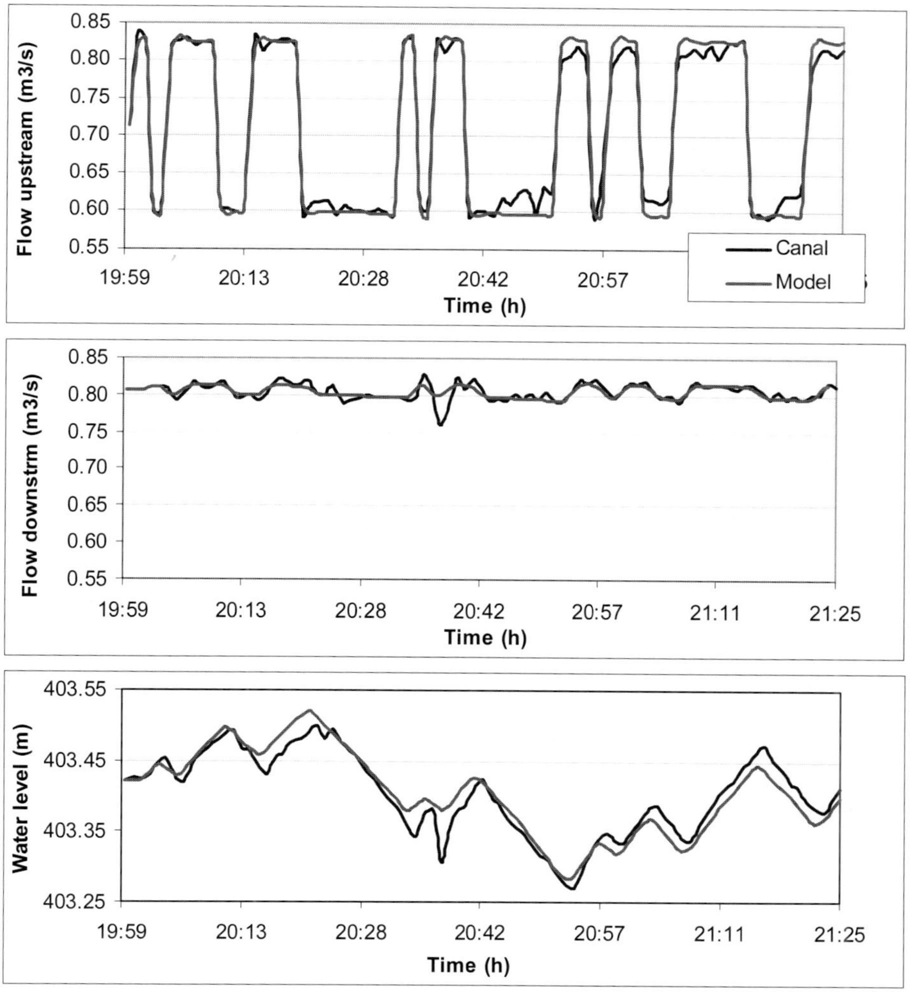

Figure 2.10 Identification test on actual and modeled canal reach

A way to identify the critical resonance without identifying the entire frequency spectrum is to use proportional feedback and to increase the gain factor until instability occurs. Schuurmans, J. (1997) and more recent Burt (2005) has shown that this is an efficient way to find the limitation of the controller gains, especially as all dynamics within the control loop such as measurement, filtering and flow controller effecting the control action, are part of the closed loop. Figure 2.11 presents the procedure to find the resonance peak and frequency of the critical wave that is in counter phase with the control action. The test is done with the flat demonstration canal reach. First, the proportional feedback gain is set to K_p=-180. The closed loop controlled reach is unstable. Next, a gain of K_p=-100 is used. Now the oscillation in the water level dampens out. Finally, a gain of K_p=-114 is applied to make the water level oscillate with a wave that has a constant amplitude. The system is semi-stable and oscillates with the frequency of the basic frequency. With this gain value, the open loop controlled water system will have a magnitude of 1 and be in counter phase (phase lag = -180°) with the control action. The frequency ω is estimated from the graph as 1.61e-2 rad/s which is in close correspondence with the other frequency estimates of the basic frequency, as previously derived. The peak of the basic frequency can be estimated with $M_{p1} = 1/Kp$ = 8.8e-3, while the peak in Figure 2.6 is 2.66e-2. The difference between these values is caused by the higher damping in the closed loop test, as here all parts of the control system, such as zero-order-hold sampling (Williamson (1991)), the flow controllers at both sides of the reach and the dynamics of the structure that execute the control actions, are identified in one time. The test in Figure 2.6 only shows the transfer function of the canal reach itself. This mismatch shows the importance of identifying critical parameters with tests that include the dynamics of all parts of the control loop.
The critical resonance does not always have to be the basic frequency. Instead, the critical resonance is the standing wave that is in counter phase with the control action. In case local control is applied, the critical resonance is one of the higher harmonics. Note that the transfer function as presented in Figure 2.6, changes for that case. For local downstream control, the transfer function from upstream inflow to the local downstream water level directly behind the structure must be considered, instead of the distant downstream water at the downstream side of the canal reach. This transfer function will show a phase lag of -180° at the critical resonance (higher harmonic) frequency.

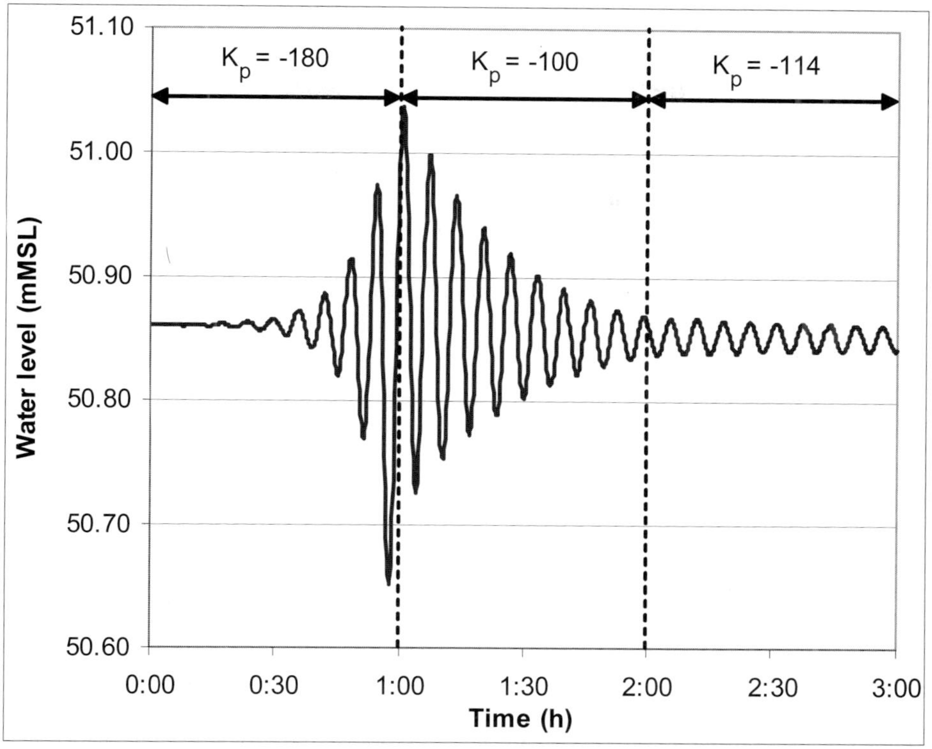

Figure 2.11 Identification of critical resonance wave with closed loop test

2.3 Structures in canals

Structures are man-made constructions that are meant to manipulate the water flows in a water system. Some structures, such as weirs and culverts, are fixed. They are designed with procedures based on steady state solutions of the water flow. Other structures, such as overshot gates, undershot gates and pumps can have different settings. They are operated to manipulate the water flows such, that the water system will go into some desired state. These adjustable structures are necessary to execute the control actions calculated by the controller. For that reason, only these structures are considered in this research.
In model based controllers the structures need to be modeled as part of the entire water system. In general, structures are modeled by an analytical formula that describes the flow through the structure as a function of the water levels at both sides of the structure and the dimensions, such as width, adjustable gate height and a calibration coefficient. This calibration coefficient describes the way the stream lines of the flow enter and leave the structure. The entire set of variables and parameters is used in submerged structures (S4 in Figure 2.2). Structures that have a low downstream water level are free flowing (S3, S4 and S6 in Figure 2.2). In that case the downstream water level is not influencing the structure flow.

The calibration coefficient is hard to estimate. For controllers that contain a feedback part, such as Proportional Integral controllers or Model Predictive Controllers a modeled structure that is dissonant compared to the actual structure, does not pose a big problem. This is because the controller will continue to adjust the gate until the objective of the management is met. For calculating the volume that passes a structure, for example an offtake undershot gate to a farmer, the inaccurate formula is a problem though. If the gate is not calibrated on a regular basis, these miscalculations can be more than ten percent per daily volume. A way to measure the flow with an inexpensive measurement object is to use a measurement flume (Bos (1989), Clemmens et al. (2001)).

2.3.1 Overshot gates

An overshot gate is a structure that backs up water, as the water has to flow over the crest of the gate. The spillway S6 at the end of the canal in Figure 2.2 is an overshot gate. Figure 2.12 shows a free flowing overshot gate, while Formula 2.7 gives the flow formula of the structure (Bos (1989), Brouwer (2004)). In this research the formula of a broad crested weir is used that is based on Bernoulli's equation describing the conservation of energy. The energy at the upstream side has two components; the potential energy of the water level above the crest and the kinetic energy in the velocity of flow. For two reasons, the kinetic energy at upstream side is generally neglected. First of all, the wetted area upstream of the structure is large, so the velocity is still low. The resulting kinetic energy is usually only a few percent compared to the potential energy. Second, the velocity is not measured in the canal reaches that are considered in this research. In case the velocities are high and can not be neglected, the kinetic energy can be calculated by an iterative procedure and can then be accounted for in the flow formula.

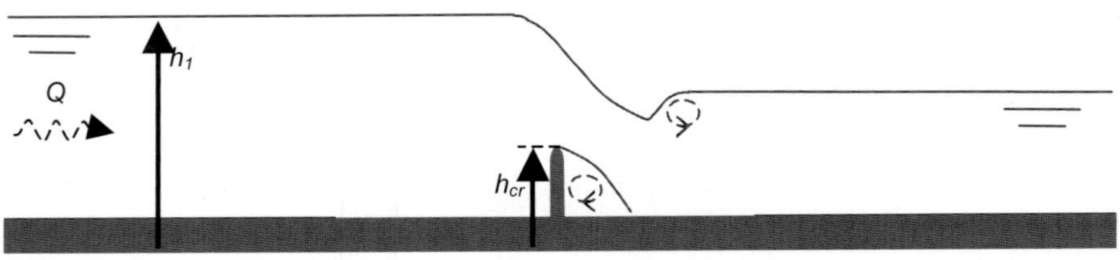

Figure 2.12 Free flowing overshot gate

$$Q(k) = C_g \cdot \frac{2}{3} \cdot W_g \cdot \sqrt{\frac{2}{3} \cdot g} \cdot \left(h_1(k) - h_{cr}(k)\right)^{\frac{3}{2}}$$ Formula 2.7

where Q represents the flow over the structure (m^3/s), C_g the calibration coefficient, W_g the width of the gate (m), g the gravitational acceleration (=9.81 m/s^2), h_1 the upstream water level (mMSL), h_{cr} the crest level (mMSL) and k the time step index.
The flow can be decreased by raising the gate and vice versa. The upstream water level can easily be controlled by an overshot gate. This can be seen from the power of 3 over 2 of the upstream water depth above the crest in the structure formula. In case the upstream water level rises due to a disturbance inflow, the flow over the structure will increase more than proportional, bringing the water level down again. This can be considered as a natural feedback of the structure on the upstream water level. A disadvantage of the overshot gate is that sediment will accumulate at the bottom just in front of the structure as the velocity of the water decreases there.
When the structure has to be used in a simplified, linearized model, the flow Formula 2.7 has to be linearized resulting in Formula 2.8 (Bosgra (2003)):

$$Q(k+1) = Q(k) +$$

$$C_g \cdot W_g \cdot \sqrt{\frac{2}{3} \cdot g \cdot \left(h_1(k) - h_{cr}(k)\right)} \cdot \Delta h_1(k) - \qquad \text{Formula 2.8}$$

$$C_g \cdot W_g \cdot \sqrt{\frac{2}{3} \cdot g \cdot \left(h_1(k) - h_{cr}(k)\right)} \cdot \Delta h_{cr}(k)$$

The overshot gate is usually operated by electric motors that move the gate with a mechanical transmission or a steel cable. The velocity of the gate movement is limited by the maximum power of the motor and the gearing. Another constraint of this structure is the minimum and maximum position of the gate crest. To limit frequent or continual gate movements, small control actions are usually summed in the memory of the controller without moving the gate. The motor is only engaged if the sum exceeds a certain dead band (e.g. 3 mm).

2.3.2 Undershot gates

Undershot gates have a gate that is put into the water from the top down. The water flows under the gate. The stream lines of the upper part of the flow, just before the gate, bend down to pass the gate opening, causing the actual flow opening to be contracted. Usually, a value is found of 0.63 by which the flow opening is contracted compared to the gate opening.
The flow through the undershot gate can be free or submerged. The head gate S3 and undershot gate S4 in Figure 2.2 are free flowing, while undershot gate S5 is submerged. Figure 2.13 presents a free flowing and submerged undershot gate, while Formula 2.9 and 2.10 give the structure flows (Bos (1989), Brouwer (2004)).

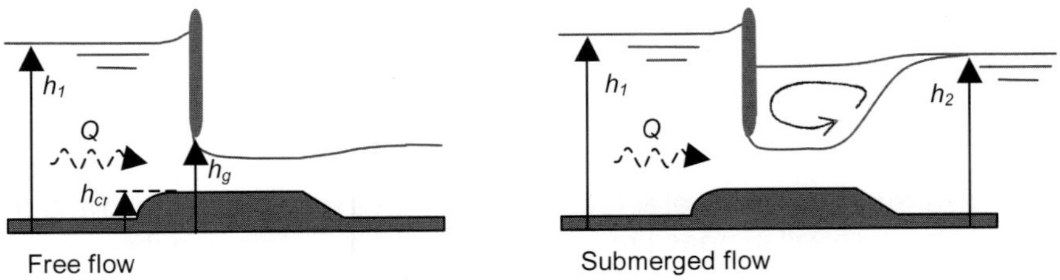

Figure 2.13 Free flowing and submerged undershot gate

Structure flow through a free flowing undershot gate:

$$Q(k) = C_g \cdot W_g \cdot \mu_g \cdot (h_g(k) - h_{cr}) \cdot \sqrt{2 \cdot g \cdot \left(h_1(k) - h_{cr} + \mu \cdot \left(h_g(k) - h_{cr}\right)\right)}$$

Formula 2.9

Structure flow through a submerged undershot gate

$$Q(k) = C_g \cdot W_g \cdot \mu_g \cdot \left(h_g(k) - h_{cr}\right) \cdot \sqrt{2 \cdot g \cdot \left(h_1(k) - h_2(k)\right)}$$

Formula 2.10

where Q represents the flow through the structure (m^3/s), C_g the calibration coefficient, W_g the width of the gate (m), μ_g the contraction coefficient, h_1 the upstream water level (mMSL), h_2 the downstream water level (mMSL), h_g the gate height (mMSL), h_{cr} the crest level (mMSL), g the gravitational acceleration (=9.81 m/s^2) and k the time step index.

In general, an undershot gate is free flowing when the downstream water level is lower than the bottom of the gate and drowned when it is higher. In this research, the level of the bottom of the gate is referred to as gate height. Due to the square root of the water level head in the formulas, undershot gates are not as well suited as overshot gates to control the upstream water level, but for controlling the downstream water level by changing the flow, they perform better. By changing the gate height, the flow can be set more precisely than by using an overshot gate.

Formula 2.11 and 2.12 give the linearized, discretized flow equations:

Linearized structure flow through a free flowing undershot gate:

$$Q(k+1)=Q(k)+\frac{g\cdot C_g\cdot W_g\cdot \mu_g\cdot(h_g-h_{cr})}{\sqrt{2\cdot g\cdot\left(h_1(k)-\left(h_{cr}+\mu_g\cdot\left(h_g(k)-h_{cr}\right)\right)\right)}}\cdot\Delta h_1(k)+\left(\begin{array}{l}C_g\cdot W_g\cdot\mu_g\cdot\sqrt{2\cdot g\cdot\left(h_1(k)-\left(h_{cr}+\mu_g\cdot\left(h_g(k)-h_{cr}\right)\right)\right)}-\\ \dfrac{g\cdot C_g\cdot W_g\cdot\mu_g^{2}\cdot(h_g-h_{cr})}{\sqrt{2\cdot g\cdot\left(h_1(k)-\left(h_{cr}+\mu_g\cdot\left(h_g(k)-h_{cr}\right)\right)\right)}}\end{array}\right)\cdot\Delta h_g(k)$$ Formula 2.11

Linearized structure flow through a submerged undershot gate:

$$Q(k+1)=Q(k)+\frac{g\cdot C_g\cdot W_g\cdot\mu_g\cdot\left(h_g(k)-h_{cr}\right)}{\sqrt{2\cdot g\cdot\left(h_1(k)-h_2(k)\right)}}\cdot\Delta h_1(k)-\frac{g\cdot C_g\cdot W_g\cdot\mu_g\cdot\left(h_g(k)-h_{cr}\right)}{\sqrt{2\cdot g\cdot\left(h_1(k)-h_2(k)\right)}}\cdot\Delta h_2(k)+C_g\cdot W_g\cdot\mu_g\cdot\sqrt{2\cdot g\cdot\left(h_1(k)-h_2(k)\right)}\cdot\Delta h_g(k)$$ Formula 2.12

Basically, the constraints of the overshot gate apply to the undershot gate. The gate can be fully closed and fully opened. The difference with the overshot gate is that when an undershot gate is fully closed, the flow is zero, while in case of an overshot gate that does not have to be the case. In case the overshot gate is fully closed at its highest level, the water can still flow over the crest, so the flow is not fully stopped. As with the overshot gate, there is a constraint on the minimum and the maximum change in gate height.

2.3.3 Pumps

Pumps (S1 in Figure 2.1) are electro-mechanical motors that lift water by means of rotating blades or a screw propeller. The maximum discharge capacity that a pump can lift depends on the maximum power of the electrical motor and the water pressure against the outflow of the pump. This pressure is a function of the (negative) head between upstream water level and downstream water level. Given the maximum power that a motor can use, the flow becomes less when the (negative) head becomes higher. In case a pump lifts water to tidal water such as the sea, the pumped volume given for a fixed amount of energy consumption, is

higher just before and after a low tide, compared to the pumped volume during high tide. Pumping during low tide is usually not possible, because of the construction of the blades and, besides that, is not cost effective. During low tide, large volumes of water can be discharged through gravity driven structures without energy consumption.
In general, the operation of a pump is very straightforward. The pump is turned on and starts running at full capacity when the upstream water level becomes higher than a certain switch on level. The pump stops again when the water has dropped lower than a certain switch off level. In between switching off and on, there exists a dead band. Nowadays, more pumps are designed with variable frequency drives that can run at variable speeds. By using a more complex local flow controller, these pumps can be set to any flow between zero and maximum capacity.
All pumps have as minimum constraint zero flow and as maximum constraint the maximum capacity as limitations on the pumped flow. The simple pumps have as extra constraint that they can only be turned either off or on. For these type of pumps it is also custom that once switched to the off or on status, they have to remain in that status for a certain time period to avoid wear and tear. This wear is due to the extra acceleration and deceleration forces during the switch over.

2.4 Boundary conditions on canals

The systems examined in this research consist of canal reaches and structures in these canals. This water system is influenced from outside by the boundary conditions. In a drainage system as depicted in Figure 2.1, the boundary conditions are the rainfall-runoff at the upstream side and the tidal water level at the downstream side of the canal. In an irrigation system as shown in Figure 2.2, these conditions are the upstream water level in the feeder reach and the offtake flows in each canal reach. Note that the downstream water level is not a boundary condition. The spillway is free flowing and so the downstream water level does not influence the water system.
When the boundary conditions have a constant value, the water system will reach a steady state after some time. In case the regular water management actions are manually or automatically executed, the steady state will show controlled water levels at setpoint. Only if a boundary condition changes, the water system will be disturbed and the control system has to respond on this by correcting for water levels that are off setpoint. This means that from a control point of view, disturbances can be seen as changing boundary conditions e.g. run-off that changes with the course of the storm event, a sea water level tide and changes in offtake flows.
When a disturbance can be predicted as a signal over the period ahead, this signal can be used in a feedforward controller to improve the water management. These disturbances are referred to as known disturbances. Other disturbances can not be measured or predicted in advance and are therefore called unknown disturbances. Examples of unknown disturbances are unpredicted rainfall and farmers that take more water out of the canal than recorded in the offtake schedule.

A model based controller that efficiently minimizes the effect of a known disturbance on the water levels in the system needs to use a sufficiently long prediction time horizon. In order to have time to counteract a disturbance, the controller needs to include a relevant disturbance in advance.

2.4.1 Rainfall-Runoff

The rainfall-runoff (D1 in Figure 2.1) can be predicted from the forecast of the precipitation and a hydrologic model. The forecast of the precipitation for the long term is computed with models of the lower part of the atmosphere. The short term forecast can be improved by using radar images of the clouds. Finally, the measurements of rainfall on the ground can be used to record the present precipitation.
The forecasted precipitation is used as input for a hydrologic model. The hydrologic model contains the processes for infiltration into the ground water, storage of water in the soil, seepage from groundwater to open water, storage on the land, direct runoff from the land when the infiltration capacity is exceeded and other relevant hydrological processes. Singh & Woolhiser (2002) give a thorough summary of the world-wide used hydrological models that link precipitation to rainfall-runoff flow into the open water system. The result of the forecasted precipitation and the hydrologic model is a prediction of the runoff flow into the canal over the period ahead e.g. over 3 days ahead.

2.4.2 Tidal water level

The water level tide (D2 in Figure 2.1) can be predicted from the orbits of the sun, earth and moon, years ahead. With the forecast of the wind force and direction for the next days, these predictions can be improved. This results in a prediction of the tide e.g. 24 hours ahead which, in The Netherlands, is approximately 2 cycles of low and high tide.

2.4.3 Offtake flows

Farmers in some irrigation districts can take water on-demand, without prior notice. This method of water supply only functions for canals with short delay times in the canal reaches a large amount of storage in the canal and sufficient water available at the upstream source. More often, farmers need to contact the irrigation district to order water some time in advance. Their demands are recorded in an offtake schedule (D3 in Figure 2.2). In order to cope with conflicting demands, an operator communicates back to the farmers shifts in timing of taking the water. Depending on the delay time from the upstream source to the last offtake, the minimum time for ordering in advance can be determined. In general, this is a couple of hours in advance.

2.5 Conclusions on open water systems

In this chapter the sub-systems of open water systems that are part of the closed loop or influence the water system at its boundaries are analyzed. The sub-systems in the closed loop are the water level dynamics in the canals and the various types of structures by which the water flows can be manipulated. Changing boundary conditions, such as run-off from precipitation, sea tide and variable offtake flows, disturb the water system and make it drift away from the desired steady state. The disturbance can be unknown or known. Known disturbances can be used to improve the performance of the controlled water system. All sub-systems can be formalized in mathematical models that can be used in model based controllers.

The most important conclusion is that the required detail of the models depends on the expected impact of each characteristic of the actual system on the stability and the performance of the water system controlled in closed loop. Long, steep canal reaches, for example, can be modeled by their most important characteristic, the delay time of waves that travel from the upstream to the downstream side of the canal reach. On the contrary, short, flat canal reaches are more sensitive to resonance waves traveling up and down the reach and may require a more detailed model using the De Saint Venant equations.

3 Model Predictive Control

Management of modern water systems requires more advanced control methods than classic feedback and feedforward. This is due to two main reasons. First, the constraints of the control structures limit the performance of the present controlled water system. For example, pumps that pump water out of a drainage system have a limited capacity, so the available storage in the water system needs to be utilized to the utmost by the controller on the pump. Second, the demands on the flexibility of a controlled water system increase over time due to socio-economic development. These developments can be e.g. more costly damage caused by inundation of the more costly modern infrastructure and land-use or higher fluctuations in water demands necessary for modern agricultural practice. Farmers in irrigation districts for example, were used to taking the water that was available to them at any time of the day. Nowadays, farmers want to receive the amount that they need at the time that is convenient to them.

Model Predictive Control (Mosca (1995), Camacho & Bordons (1999), Rawlings (2000), Overloop et al. (2003b)) is a control methodology which can apply feedback and feedforward control and use the constraints on structures and water levels in a systematic way. The fundamental differences between the control methods are illustrated in Figure 3.1. Here, the pumped flows and water levels in the storage canals of a control application on a drainage system are shown. This application is described in detail in Paragraph 6.1. All pumps together have a maximum capacity of 60 m^3/s. The disturbance inflow due to runoff from an extreme storm event is higher than this capacity plus the storage capacity between setpoint and the maximum allowed water level.

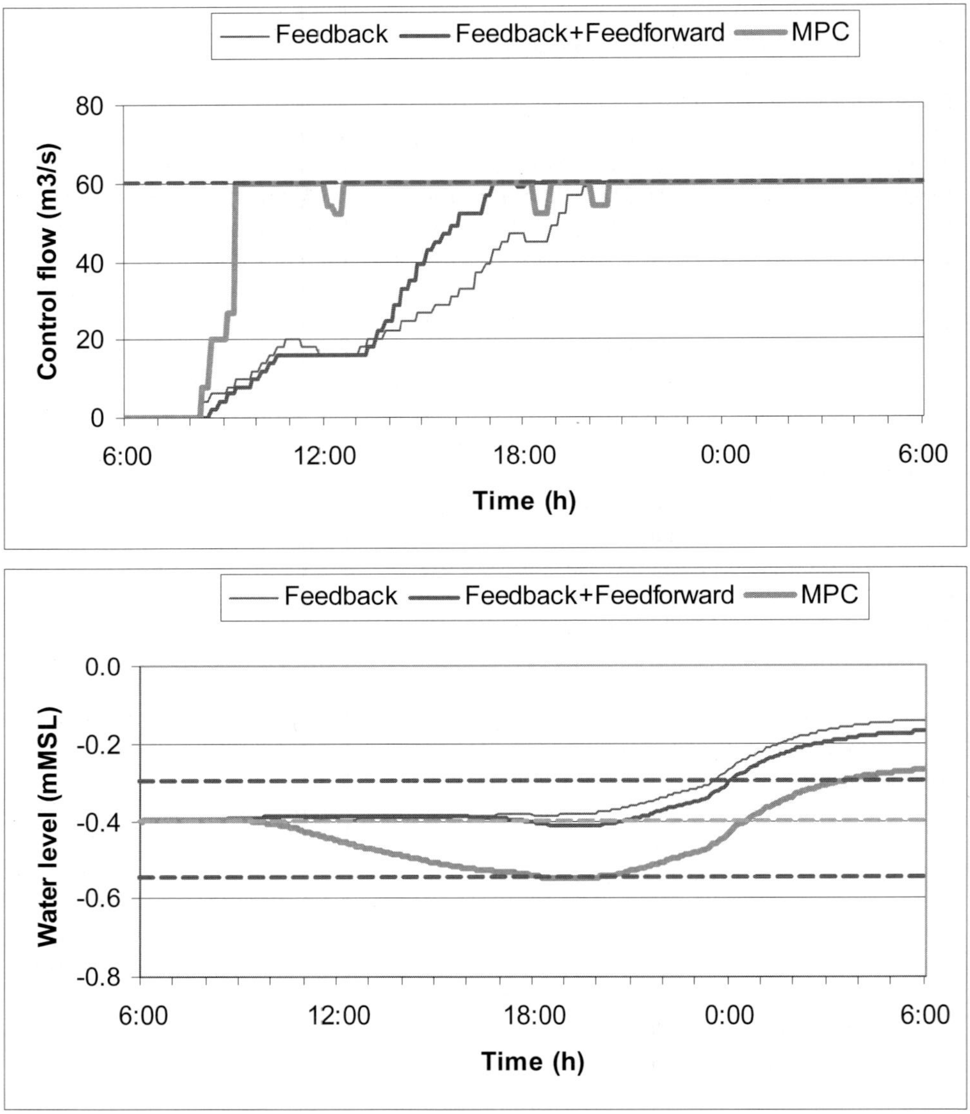

Figure 3.1 Results of different control methods

The feedback controller reacts when the disturbance causes the water level to rise. As this controller only reacts after a deviation occurs, its reaction is always late. Once the control flow becomes higher than the maximum capacity of the pumps, the outflow is limited and eventually, the water level rises much higher than the maximum allowed water level.

The feedback controller which is extended with a feedforward controller uses the prediction of the effect that the disturbance has on the water level and

counteracts on this, in order to keep the water level exactly to zero. This works well until the control flow exceeds the maximum pump capacity. Also with this control method, the maximum allowed water level constraint is violated.
The Model Predictive Controller uses the same prediction of the disturbance. It applies an objective function in which the water level deviation from setpoint is minimized over a prediction horizon. It uses the information on the constrained control flow, so it can predict the high water levels at the end of the prediction horizon. To minimize the water level deviations over the entire prediction horizon, MPC starts earlier with pumping out water to lower the water level before the disturbance inflow even takes place. In this way, the maximum allowed water level is violated to a much lower extent.
Model Predictive Control is a model based controller. These types of controllers emerged from the chemical industry in the 70's (Mosca (1995)). Boom & Backx (2001) describe the history of the evolution of the control methodology. Basically, these types of controllers are necessary, as the products made in this sector are fabricated very close to the limits of the quality specifications. This requires a control method that can work close to constraints by using information of these constraints on the process. This is very similar to the present specifications on controlled water systems.
Figure 3.2 shows a structure diagram of a water system controlled by a Model Predictive Controller. MPC contains the following components:

- Internal model. In any prediction a model is required. In controlled water systems, this model is used to predict the water levels and flows over the prediction horizon as a result of the disturbance flows and the control flows.
- Objective function. This function captures the objective of the controlled water system. The overall objective is built up from separate conflicting sub-objectives. Each sub-objective is weighted to give a relative penalty to that part of the overall objective. All weighted penalties together form the objective function. For controlled water systems, these conflicting sub-objectives can be the water level deviation from setpoint weighted against the energy consumption that is required to maintain a small deviation. An additional example is the water level deviations in adjacent canal reaches weighted against each other and against the required gate adjustments to maintain small water level deviations.
- Constraints. Constraints are limitations on the solution calculated by the Model Predictive Controller. These limitations can be of a physical nature or consist of operational specifications. In water systems the structures have limited capacities. The controller is not allowed to calculate a solution that violates these physical limitations. Also, operational specifications are imposed on the solution, such as water levels that are to remain within a certain range around setpoint or structures that are not allowed to switch more often than a certain minimum time period.
- Optimization. The objective function has to be minimized by using a numerical optimization algorithm. This optimization needs to take the constraints into account.

- Receding horizon. After each optimization, an open loop result over the prediction horizon is available. Over this horizon, the required control actions and the resulting water levels are presented. In a Decision Support System, this information is already a great help to the operators, as they can look into the future given the present predictions. Even though the control actions over the entire horizon are given, only the first action is actually implemented in closed loop. After one control time step the entire optimization is repeated. In this way, the open loop horizon recedes into the future, always using the most recent measurements and predictions.

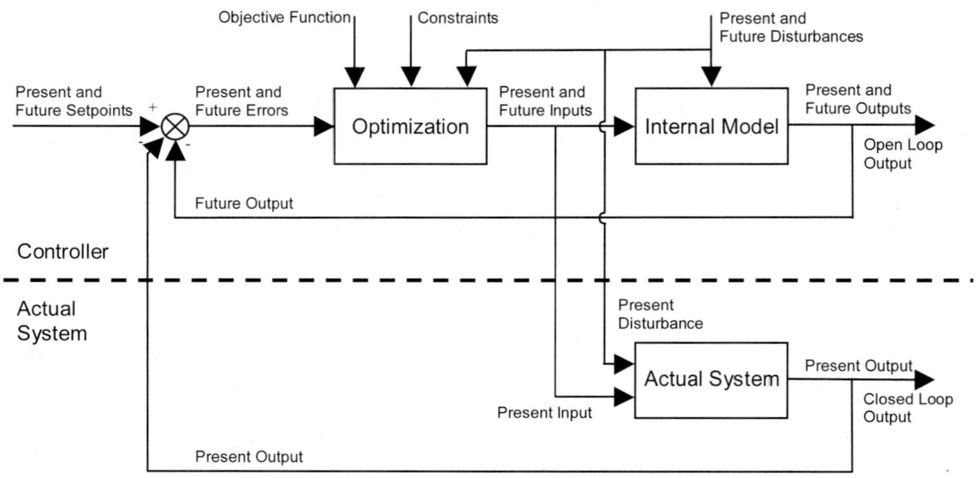

Figure 3.2 Structure diagram of Model Predictive Control on actual system

Next, all parts of the Model Predictive Controller, as configured for control of open water systems, are discussed.

3.1 Internal model

The internal model is used to predict the future states of the controlled water system. As input, it uses the present and future disturbances and the present and future control actions. These control actions are calculated by the optimization. The outputs of the model are the present and future water levels and flows. The internal model needs to be a sufficiently accurate representation of the actual system. What sufficiently accurate is exactly, is hard to define in advance. The receding horizon principle though, facilitates the permanent update of the model. This update keeps even a simple model accurate, at least for the predictions in

the near future. The internal models used in all successful MPC applications as described in Chapter 6 were simplified to a considerable extent.
Like any real system, the actual water system is non-linear. To be able to use linear algebra theory and generally available computational tools to solve the control problem, it is preferable to convert these non-linear sub-systems into linear systems. The main sources of this non-linearity of controlled water systems are:

- The changing dynamics of a canal reach in the transition from low to high flow and vice versa. These dynamics change, but not to a large extent. The reason for this small change in dynamics is that the controller keeps the controlled water levels close to setpoint. The resulting depths along the reach do not change considerably (see Figure 2.5).
- The non-linearity of the structure flows that depend on the varying water levels and the gate height as shown by Formulas 2.7, 2.8 and 2.9.

The non-linearity of the structure flows can be eliminated in case the structure flows are used as input to the internal model instead of the gate height. The closed loop implementation of the actual control action is performed by using a flow controller. This flow controller converts the required flow into a gate height by using actual measurements of the water levels and inverting the structure formula. This configuration is generally referred to as Master-Slave control (Schuurmans, J. (1997)). In the MPC application, as described in paragraph 6.2 and 6.3, such a flow controller is used to set the gate height of the controlled gate. As this research strives for a standardized formalization of Model Predictive Control for all types of open water systems, general mathematical methods are applied to build up controllers. These general methods are based on linear systems. The internal model can be linearized in each time step along the variable trajectories over the prediction horizon. This results in *n* subsequent linear models. In case there are minor transients in the water movements, this is an accurate approximation of the non-linear system over the prediction horizon. When transients are sharper, a number of iterations can be executed, in which each linearization is performed along a trajectory of the weighted new and previous variables as given in Formula 3.1 (Stelling & Booij (1994)):

$$P(k) = f\left(\xi \cdot h_j(k) + (1-\xi) \cdot h_{j-1}(k), \ldots\right) \qquad \text{Formula 3.1}$$

where *P(k)* represents the linear model of the water system at time step *k*, *h* is one of the variables, ξ the weight factor and *j* the iteration index.

Water systems consist of sub-systems that interact. Often, the main interactions are the structure flows in between the canal reaches, which are calculated by the controller. As water levels are measured at multiple locations and multiple structures are controlled, the controller requires a multivariable configuration. State space models allow for compact, multivariable formalization of linear models (Vandevegte (1990)). Many algorithms and tools for solving control problems are based on this standardized model description. The state space model as used in this research is presented in Formula 3.2

$$
\begin{aligned}
x(k+1) &= A(k)\cdot x(k) + B_u(k)\cdot u(k) + B_d(k)\cdot d(k) \\
y(k) &= C\cdot x(k)
\end{aligned}
$$

Formula 3.2

where x represent the states of the water system, A the system matrix, B_u the control input matrix, B_d the disturbance input matrix, u the inputs calculated by the controller, d the disturbances, C the output matrix, y the outputs of the modeled water system that are measured in the actual system and k the time step index. The required discretization in space is made by using appropriate states, for example the water levels along a canal reach, while the required discretization in time is build up by the n steps in time over the prediction horizon. This discretization allows for numerical solution using a computer. The simplified Integrator Delay and De Saint Venant models as described in Paragraph 2.2.2, can be written in state space. These models are used and described in detail in the MPC applications in Chapter 6.

When the state space model is extended over the prediction horizon and the initial state $x(k)$ and inputs are known, the solution can be computed directly. The part $|k$ is added to indicate that the future variables and matrices are calculated with information available at time step k. In Formula 3.3 the state space model is extended over the prediction horizon from k+1 to k+n, to show that all solutions can be expressed as a function of $x(k)$ and the control action and disturbances used over the prediction horizon:

$$
\begin{aligned}
& x(k+1\,|\,k) = A(k\,|\,k)\cdot x(k) + B_u(k\,|\,k)\cdot u(k\,|\,k) + B_d(k\,|\,k)\cdot d(k\,|\,k) \\
& x(k+2\,|\,k) = A(k+1\,|\,k)\cdot x(k+1\,|\,k) + B_u(k+1\,|\,k)\cdot u(k+1\,|\,k) + B_d(k+1\,|\,k)\cdot d(k+1\,|\,k) \\
& x(k+2\,|\,k) = \\
& A(k+1\,|\,k)\cdot A(k\,|\,k)\cdot x(k) + A(k+1\,|\,k)\cdot B_u(k\,|\,k)\cdot u(k\,|\,k) + A(k+1)\cdot B_d(k\,|\,k)\cdot d(k\,|\,k) + \\
& B_u(k+1\,|\,k)\cdot u(k+1\,|\,k) + B_d(k+1\,|\,k)\cdot d(k+1\,|\,k) \\
& \quad \vdots \\
& x(k+n\,|\,k) =
\end{aligned}
$$

Formula 3.3

In Formula 3.4, these formulas are written in a more compact form as large matrices that can be solved by fast numerical solvers:

$$\mathbf{X}=\begin{bmatrix} x(k\mid k) \\ x(k+1\mid k) \\ \vdots \\ x(k+n\mid k) \end{bmatrix} \quad \boldsymbol{U}=\begin{bmatrix} u(k\mid k) \\ u(k+1\mid k) \\ \vdots \\ u(k+n-1\mid k) \end{bmatrix} \quad \mathbf{D}=\begin{bmatrix} d(k\mid k) \\ d(k+1\mid k) \\ \vdots \\ d(k+n-1\mid k) \end{bmatrix}$$

$$\mathbf{A}=\begin{bmatrix} I \\ A(k\mid k) \\ A(k+1\mid k)\cdot A(k\mid k) \\ \vdots \\ A(k+n-1\mid k)\cdot A(k+n-1\mid k)\cdots A(k\mid k) \end{bmatrix}$$

$$\mathbf{B}_u=\begin{bmatrix} 0 & 0 & \cdots & 0 \\ B_u(k\mid k) & 0 & & 0 \\ A(k+1\mid k)\cdot B_u(k\mid k) & B_u(k+1\mid k) & & 0 \\ \vdots & & \ddots & \vdots \\ A(k+n-1\mid k)\cdot A(k+n-2\mid k)\cdots A(k+1\mid k)\cdot B_u(k\mid k) & & & B_u(k+n-1\mid k) \end{bmatrix}$$

$$\mathbf{B}_d=\begin{bmatrix} 0 & 0 & \cdots & 0 \\ B_d(k\mid k) & 0 & & 0 \\ A(k+1\mid k)\cdot B_d(k\mid k) & B_d(k+1\mid k) & & 0 \\ \vdots & & \ddots & \vdots \\ A(k+n-1\mid k)\cdot A(k+n-2\mid k)\cdots A(k+1\mid k)\cdot B_d(k\mid k) & & & B_d(k+n-1\mid k) \end{bmatrix}$$

$$\mathbf{C}=\begin{bmatrix} C & \cdots & 0 \\ \vdots & C & \\ & & \ddots & \\ 0 & & C \end{bmatrix}$$

$$\mathbf{X}=\mathbf{A}\cdot x(k)+\mathbf{B}_u\cdot\boldsymbol{U}+\mathbf{B}_d\cdot\mathbf{D}$$
$$\boldsymbol{Y}=\boldsymbol{C}\cdot\boldsymbol{X}$$ Formula 3.4

The matrix ***D*** represents a prediction of the disturbances over the prediction horizon. In Paragraph 2.4 the common disturbances on open water systems, such as water level tide, rainfall runoff and offtake schedules are discussed. The matrix ***U*** represents the result of the optimization as described in Paragraph 3.4.
The initial condition *x(k)* needs to be estimated at the beginning of every control time step. A common way of estimating the states of a model is by using a state estimator:

$$x(k)=x(k\mid k-1)+\boldsymbol{L}\cdot\left(y^{*}-C\cdot x\left(k\mid k-1\right)\right)$$ Formula 3.5

where *x(k|k-1)* represents the present state as computed at the previous time step, ***L*** the state estimator, y^* the measured output of the actual water system.

The state estimator can be computed using the well-known Kalman theory (Kalman (1960), Reddy (1994)). In the research described in this dissertation, estimates of measurement noise and model noise are taken from expert judgment without doing any stochastic experiments. The state estimator feeds back the difference between measured states and computed states. For measured states this estimator feedback has a relative high gain (correcting more towards the value of the measurement), while for unmeasured states the feedback gain depends on the extent of coherence between the state and the measured state. In the MPC-application described in Paragraph 6.2, a Kalman filter is implemented.

3.2 Objective function

The objective function formalizes the goals that the controller has to try to achieve. The objective function contains sub-objective that are added up. These sub-objectives can be conflicting. When a relative penalty is given to each of these sub-objectives to indicate the relative importance of the sub-objective, a quantified and dispassionate solution can be computed by minimizing the objective function.

The controller has to bring the state of the water system into some desired state. In case this desired state is considered constant over the prediction horizon, all states over the prediction horizon $\boldsymbol{X}$ are part of the objective function. In general, bringing the state of the water system back to the desired state has to be performed with restrictions on the input that controls the water system. A minimum number of adjustments may be used or as little energy consumption as possible. Consequently, the input over the prediction horizon $\boldsymbol{U}$ is also part of the objective function. The relative penalties on the states of the water system are put into the weight matrix Q that is multiplied by the state vector x. Over the prediction horizon n this weight matrix becomes $\boldsymbol{Q}$ with n times Q on the diagonal. The relative weight on the input is put into weight matrix R. Over the prediction horizon n this weight matrix becomes $\boldsymbol{R}$ with n times R on the diagonal.

The objective function in this research is set up by using Quadratic Programming. The reason is, that using the square of the states and inputs, penalizes both positive and negative deviations and higher absolute deviations are penalized more than proportional due to the power of 2. An additional advantage is the easy computation of the derivative of the objective function given as a Quadratic Programming problem. The minimum of the objective function can then be found by making the derivative equal to zero. Ultimately, the finite horizon objective function over the prediction horizon n used in this research is:

$$\min_{u} J = \boldsymbol{X}^T \cdot \boldsymbol{Q} \cdot \boldsymbol{X} + \boldsymbol{U}^T \cdot \boldsymbol{R} \cdot \boldsymbol{U} \quad \text{Formula 3.6}$$

As open water systems are inherently stable systems, there is no need to add an infinite horizon part into the objective function, especially as long prediction horizons are already used and the operators are only interested in this upcoming part of the future. Formula 3.7 presents the detailed formalization of the objective function, which is used in most of the MPC applications described in Chapter 6. This formula shows that the objective function can become a large multi-objective optimization problem, when the prediction horizon n and the number of sub-systems m are large.

$$\min_{\Delta u} J =$$

$$\sum_{i=0}^{n}\sum_{j=1}^{m}\{e_j(k+i\,|\,k)\cdot Q_{e,j}\cdot e_j(k+i\,|\,k)\}+$$
$$\sum_{i=0}^{n}\sum_{j=1}^{m}\{\Delta e_j(k+i\,|\,k)\cdot Q_{\Delta e,j}\cdot \Delta e_j(k+i\,|\,k)\}+ \qquad \text{Formula 3.7}$$
$$\sum_{i=0}^{n-1}\sum_{j=1}^{l}\{\Delta u_j(k+i\,|\,k)\cdot R_{\Delta u,j}\cdot \Delta u_j(k+i\,|\,k)\}$$

where J represents the objective function that needs to be minimized, n the number of steps over the prediction horizon, m the number of canal reaches, l the number of structures, e_j the water level deviation from setpoint in reach j, $Q_{e,j}$ the penalty on this error, Δe_j the change in water level deviation in reach j, $Q_{\Delta e,j}$ the penalty on this change in error, Δu_j the change in gate height or pump flow of structure j and $R_{\Delta u}$ the penalty on this change in structure setting. The change in error and the change in structure setting are defined as follows:

$$\Delta e(k)=e(k)-e(k-1) \qquad \text{Formula 3.8}$$

$$\Delta u(k)=u(k)-u(k-1) \qquad \text{Formula 3.9}$$

This selection of variables in the objective function, including the change in error and the integral of this variable, the error itself, results in control actions, that bring the water levels back to the desired levels without static deviation.
The weight matrices Q and R are used to give relative penalties to the variables. Using these penalties, the behavior of the controlled water system can be shaped. High penalties on the water level deviation relative to low penalties on the gate height changes results in tight control. There is a risk of destabilizing the actual controlled system, though. This is due to un-modeled dynamics of the actual water system or model mismatches. As these dynamics are not known by the controller, it can literally not control them. A more sensible approach is to use a realistic estimate of how much a state or input may vary. This estimate is captured in a Maximum Allowed Value Estimate (*MAVE*). For water level deviations, this value can be taken from the minimum and maximum allowed water levels relative to the setpoint. For gate height changes, the maximum gate height change can be used, while for pumps the maximum pump capacity is a logical value. In case the entries in Q and R, which are multiplied by the various

variables that need to be weighted, are set to the reciprocal of the square of the *MAVE* of that variable, the sub-objectives in the objective function are normalized. Formula 3.7 now becomes:

$$\min_{\Delta u} J =$$

$$\sum_{i=0}^{n}\sum_{j=1}^{m}\left\{e_j(k+i|k)\cdot\frac{1}{e_{MAVE,j}^2}\cdot e_j(k+i|k)\right\}+$$

$$\sum_{i=0}^{n}\sum_{j=1}^{m}\left\{\Delta e_j(k+i|k)\cdot\frac{1}{\Delta e_{MAVE,j}^2}\cdot \Delta e_j(k+i|k)\right\}+ \qquad \text{Formula 3.10}$$

$$\sum_{i=0}^{n-1}\sum_{j=1}^{l}\left\{\Delta u_j(k+i|k)\cdot\frac{1}{\Delta u_{MAVE,j}^2}\cdot \Delta u_j(k+i|k)\right\}$$

where $e_{MAVE,j}$ represents the Maximum Allowed Value Estimate of the water level deviation in reach *j*, $\Delta e_{MAVE,j}$ the Maximum Allowed Value Estimate of the change in water level deviation in reach *j* and $\Delta u_{MAVE,j}$ the Maximum Allowed Value Estimate of the change in structure setting of structure *j*. Now, a calculated maximum gate height change or maximum pump flow has the same penalty as a maximum water level deviation that needs to be corrected.
Priorities between similar sub-objectives can easily be incorporated. When, for example, the freeboard in one canal reach is smaller than the other reaches, the e_{MAVE} of this reach can be selected smaller, resulting in tighter control for this specific reach. A priority of one structure to be used first compared to other structures, can be set by using a higher Δu_{MAVE} for this structure.
When the objective function is set up as previously described, it can be used directly as total performance indicator on the closed loop results of the controlled water system. In case the results of various controllers are compared with the standard performance indicators as described in Schuurmans, W. et al. (1995), there is often a problem concerning the mutual importance of the performance indicators. One controller can have a lower maximum absolute error (MAE), while another has a lower integrated average absolute gate movement (IAW). The question remains, which controller performs better? Using the objective function over the entire result time frame, allows for an objective and total judgment of the performance, as all relevant performance indicators are part of the objective function.

3.3 Constraints

Constraints are the physical and operational limitations on the controlled water system. The optimization algorithm used in the Model Predictive Controller is able to narrow its solution area by taking these constraints into account. Some constraints may never be violated by the solution, as they represent fixed values, such as the maximum pump capacity or maximum gate height change. In case no other safety restrictions are implemented in the structure itself, violation of these values may damage the equipment. These constraints are hard constraints. Other constraints are less rigid and may be violated to some extent. These constraints are soft constraints (Boom & Backx (2001), Hovland (2004)). The difference in implementation is that hard constraints are put into the optimization as a hard limitation to a state or input, while soft constraints are implemented as extra penalties when the state or input violates its limitation. Formula 3.11 formalizes hard constraints on state and input:

$$\begin{aligned} E \cdot x(k) &\leq x_{\lim}(k) \\ F \cdot u(k) &\leq u_{\lim}(k) \end{aligned}$$

(Each row contains one inequality) Formula 3.11

where E and F represent selection matrices with values 1 or -1 at the entries which are multiplied by the state or input that needs to be limited. The value -1 is used to implement constraints on state and inputs which need to be higher than a minimum value.

Soft constraints are implemented by using a virtual input signal and a virtual state. The virtual state is computed by subtracting the virtual input from the state that needs to be constrained. In the objective function there is a high penalty on the virtual state and a very low penalty on the virtual input. The result is that the virtual state is set to zero by the optimization. Now, a hard constraint is applied to the virtual input that equals the required constraint on the virtual state. The result is that the virtual state is zero when the state that needs to be constrained is within its limitation. In case a violation occurs, this virtual state is equal to the part of the state that is outside of its constraint. When a considerably higher penalty is given to the part of the state outside of its limitation, the optimization will try to avoid the violation. In Boom & Backx (2001) this implementation of soft constraints is referred to as zone performance index. Formula 3.12 gives the mathematical formalization as used in the MPC applications described in Paragraph 6.1 and 6.4:

$$\min_{\Delta Q_c, u^*} J =$$

$$\sum_{i=0}^{n} \left\{ e(k+i|k)^T \cdot Q_e \cdot e(k+i|k) \right\} +$$

$$\sum_{i=0}^{n} \left\{ \Delta e(k+i|k)^T \cdot Q_{\Delta e} \cdot \Delta e(k+i|k) \right\} +$$

$$\sum_{i=0}^{n} \left\{ e^*(k+i|k)^T \cdot Q_{e^*} \cdot e^*(k+i|k) \right\} + \qquad \text{Formula 3.12}$$

$$\sum_{i=0}^{n} \left\{ u^*(k+i|k)^T \cdot R_{u^*} \cdot u^*(k+i|k) \right\} +$$

$$\sum_{i=0}^{n-1} \left\{ \Delta Q_c(k+i|k)^T \cdot R_{\Delta Q_c} \cdot \Delta Q_c(k+i|k) \right\}$$

where e^* represents the virtual state, u^* the virtual input, e_{max} the maximum allowed value of e and e_{min} the minimum allowed value of e. Now, the penalty Q_{e^*} on the virtual state is set much higher than Q_e and R_{u^*} is taken very small. Figure 3.3 illustrates a typical trajectory of e and e*, when a violation of the soft constraint occurs.

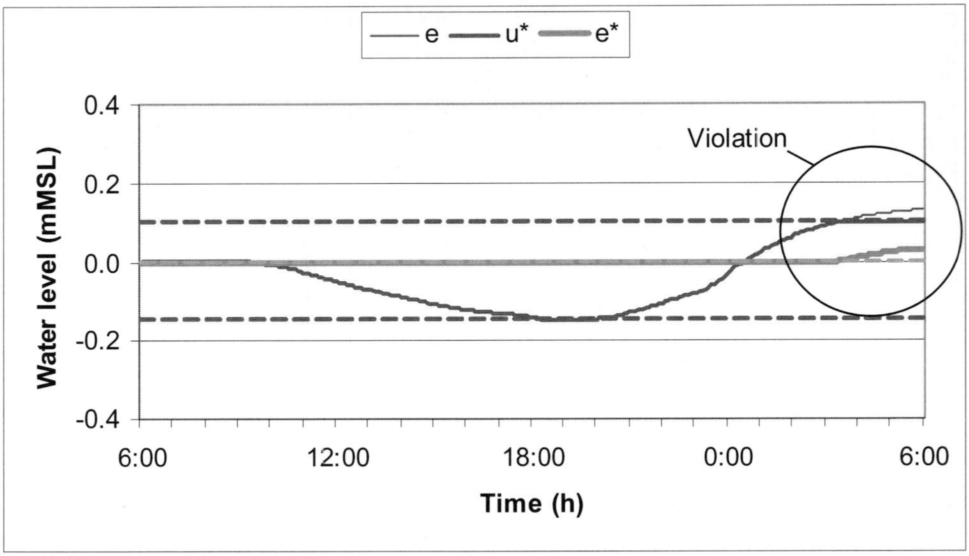

Figure 3.3 Variables used in a soft constraint

The reason soft constraints are preferred over hard constraints is that they guarantee a solution of the optimization. The use of multiple hard constraints can lead to infeasibility of the solution. In case the capacity of a structure has limited drain capacity and the water levels must remain lower than a certain maximum level, there can be a moment in time that these hard constraints conflict with one another. In that case, no solution can be calculated in which the drain capacity is not exceeded and the water level remains lower than the maximum level. The optimization algorithm will terminate with an infeasible solution.
Until now, the constraints are considered constant over the prediction horizon. Taking non-linearities out of the internal model, sometimes results in non-linear constraints. An example of this is the use of a control flow instead of the gate height as input to the internal model. This makes an internal model implemented as the Integrator Delay model linear, but makes the constraints on the flow dependant on the upstream and downstream water levels. In this research, the non-linear constraints are converted into time-variant constraints by using the previous solution of the optimization. When regular MPC is applied, this is the solution of the previous time step, shifted one step in time with the last, unknown value in the prediction horizon set equal to the previous last known value. When Model Predictive Control configured in a sequential loop as described in Chapter 4 is applied, this previous solution is the solution of the previous iteration step. This linearization is applied in the applications described in Paragraph 6.2, 6.3 and 6.5.
Another effect that makes the constraints time-variant is a time-variant disturbance that influences the constraint. A water level tide downstream of a drainage gate for example, makes the maximum flow constraint go from a very high value during low tide, to zero during high tide.

3.4 Optimization

The objective function over the prediction horizon subject to the constraints on states and inputs, needs to be minimized using an optimization algorithm. This optimization algorithm calculates the control actions over the prediction horizon that need to be implemented to get the lowest possible value of J as given in Formula 3.6. The objective function is written as a quadratic function of states and inputs.
First, an additional feedback law is introduced. This feedback law K results in a guaranteed stable controlled water system, which improves the convergence of finding the minimum solution:

$$u(k) = -K(k) \cdot x(k)$$ Formula 3.13

The time step index is added to K to make this description generally applicable for stepwise linearized internal models. The feedback law K is found by solving the well-known Linear Quadratic Regulator problem (Vandevegte (1990)) with a numerical algorithm:

$$\min_{K} J = \sum_{i=0}^{\infty} \left\{ x(k+i \mid k)^T \cdot Q \cdot x(k+i \mid k) + u(k+i \mid k) \cdot R \cdot u(k+i \mid k) \right\}$$

Formula 3.14

This objective function closely resembles the objective function of the Model Predictive Controller, especially as the weight matrices *Q* and *R* are set the same as used in MPC. An advantage of calculating this feedback law is that it can be used when the MPC solution is infeasible because of conflicting (hard) constraints. In that case, the feedback law can be applied, which will try to bring the system back to setpoint. Of course the resulting control actions are limited in order not to exceed the input constraints. Now, the stable matrix A_c is defined which is substituted in all matrices as given in Paragraph 3.1:

$$A_c(k) = A(k) - B_u(k) \cdot K(k)$$

Formula 3.15

Formula 3.16 presents the feedback law that is extended over the prediction horizon *n*:

$$\mathbf{K} = \begin{bmatrix} K(k) & \cdots & & 0 \\ \vdots & K(k+1) & & \\ & & \ddots & \\ 0 & & & K(k+n) \end{bmatrix}$$

Formula 3.16

The control actions now consist of a feedback part ***K·X*** and an additional control part ***Z*** that is required to achieve the minimum objective function subject to the constraints. Formula 3.4 now becomes:

$$\begin{aligned} \mathbf{X} &= \mathbf{A}_c \cdot x(k) + \mathbf{B}_u \cdot \mathbf{Z} + \mathbf{B}_d \cdot \mathbf{D} \\ \mathbf{U} &= \mathbf{K} \cdot \mathbf{X} + \mathbf{Z} \end{aligned}$$

Formula 3.17

To find the minimum of the objective function *J*, the gradient of the function has to be set to zero. The gradient of a quadratic function can be found by calculating the Hessian ***H*** and the Lagrangian ***f***:

$$J(\mathbf{Z}) = \frac{1}{2} \cdot Z^T \cdot \mathbf{H} \cdot Z + f \cdot Z + g$$

Formula 3.18

$$\frac{\partial J(\mathbf{Z})}{\partial \mathbf{Z}} = \mathbf{H} \cdot Z + f = 0$$

Formula 3.19

$$Z = -\frac{f}{H}$$ Formula 3.20

H and ***f*** can be found by substituting Formula 3.17 in the expression of *J* as presented in Formula 3.6 and rewriting the formula into the structure of Formula 3.18:

$$\min_{Z} J =$$
$$(A_c \cdot x(k) + B_u \cdot Z + B_d \cdot D)^T \cdot Q \cdot (A_c \cdot x(k) + B_u \cdot Z + B_d \cdot D) +$$
$$(K \cdot X + Z)^T \cdot R \cdot (K \cdot X + Z)$$

Formula 3.21

Below, the large matrices ***H***, ***f***, ***V*** and ***w*** are derived that are used as input for the numerical solver of the optimization problem.

$$H =$$
$$2 \cdot \left(B_u{}^T \cdot Q \cdot B_u + R \cdot K \cdot B_u + R + B_u{}^T \cdot K^T \cdot R \cdot K \cdot B_u + B_u{}^T \cdot K^T \cdot R\right)$$

Formula 3.22

$$f =$$
$$2 \cdot x(k)^T \cdot \left(A_c{}^T \cdot Q \cdot B_u + A_c{}^T \cdot K^T \cdot R \cdot K \cdot B_u + A_c{}^T \cdot K^T \cdot R\right) +$$
$$2 \cdot D^T \cdot \left(B_d{}^T \cdot Q \cdot B_u + B_d{}^T \cdot K^T \cdot R \cdot K \cdot B_u + B_d{}^T \cdot K^T \cdot R\right)$$

Formula 3.23

Next, the matrices for describing the constraints are derived. Formula 3.11 is extended over the prediction horizon:

$$E \cdot X \leq X_{lim}$$
$$F \cdot U \leq U_{lim}$$ Formula 3.24

Formula 3.17 is substituted in Formula 3.24:

$$E \cdot (A_c \cdot x(k) + B_u \cdot Z + B_d \cdot D) \leq X_{lim}$$
$$F \cdot (K \cdot X + Z) \leq U_{lim}$$ Formula 3.25

This formula can be rewritten as:

$$V \cdot Z \leq w \qquad \text{Formula 3.26}$$

$$V = \begin{bmatrix} E \cdot B_u \\ F \cdot K \cdot B_u + F \end{bmatrix} \qquad \text{Formula 3.27}$$

$$w = \begin{bmatrix} X_{lim} - E \cdot B_d \cdot D - E \cdot A_c \cdot x(k) \\ U_{lim} - F \cdot K \cdot B_u \cdot D - F \cdot K \cdot A_c \cdot x(k) \end{bmatrix} \qquad \text{Formula 3.28}$$

The optimization problem is now written in a general form. Note that the constant matrix ***g*** can be left out, as it has no influence on the minimization problem:

$$\min_{Z} J = \frac{1}{2} Z^T \cdot H \cdot Z + f^T \cdot Z \qquad \text{Formula 3.29}$$
$$V \cdot Z \leq w$$

The constraints determine if this optimization problem is feasible. Conflicting constraints, such as limits on the allowable water level error and limited structure flow capacities can result in an infeasible solution. Not a single optimization algorithm can find a solution. In case the solution is feasible, another characteristic of the optimization problem needs to be met to find a unique, global minimum of the objective function subject to constraints. The optimization problem needs to be convex. The principle of convexity is explained in Appendix A. When constraints are not depending on one another and they are characterized only by a minimum and maximum allowed value and the objective function by itself is convex (which it is as quadratic function of a linearized model), the optimization problem generally is convex. All MPC applications presented in Chapter 6 have convex optimization problems and therefore result in global minimum values for the objective functions used.
Bakker (2002) made a comparison of different algorithms to solve the optimization problem as given in Formula 3.29. The comparison is based on the following criteria:

- Computational speed. In case the computation of the solution is not faster than real-time, the controller cannot be implemented on an actual water system. Advantage of controlled water systems is that these systems have slow responses compared to industrial applications;
- Stability. The applied Model Predictive Controller must result in a stable controlled water system;
- Robustness. The controller has to be robust to modeling errors and unknown disturbances;

- Performance. The controller has to have a sufficient performance to fulfill the specifications on the controlled water system.

A number of algorithms are compared namely The active set method, Manipulating the constraints, Interior Point algorithms, Linear Programming, Multi parametric Quadratic Programming, Newton Raphson MPC. Most of these methods simplify the optimization problem to gain computational speed. These simplifications alter the problem to a certain extent, so stability, robustness and performance cannot be guaranteed in all cases. The two methods that solve the total optimization problem, the active set method and Interior Point algorithms, differ in computation speed. The order of the computational speed as a function of the prediction horizon of the active set method is $O(n^3)$, while for the Interior Point algorithms an order close to $O(n)$ is found. For a model of a certain size with a short prediction horizon, the active set method is faster, while for the same model with a long horizon an Interior Point algorithm is faster. For the open water systems considered in this research, the prediction horizon is relatively long. For water systems with long delay times, such as irrigation canals, the prediction time horizon needs to be at least as long as the highest delay time of the canal reaches. A better approach is to use the sum of the delay times of all canal reaches, as in that case the effects on the water level in the last reach from adjustments to the head gates are known. For drainage systems, the reaction of the water levels on a predicted forecast over the next hours or even days needs to be estimated, so also here, long prediction time horizons are necessary. The control time step cannot be selected too high, as the dynamics of the (basic) waves and fast changing disturbances such as water level tides, still need to be captured. This combination of long prediction time horizons and small control time steps results in long prediction horizons. For that reason, an Interior Point algorithm is used to solve the optimization problems of the MPC applications in this research. Appendix B presents the interior point algorithm used (Wright (1997)).
Table 3.1 gives the prediction horizons and the roughly estimated Central Processing Unit times on a regular personal computer fabricated in the year 2002 for the MPC applications as described in Chapter 6.

Table 3.1 Computation time MPC applications

Application	Time horizon (h)	Time step (s)	Prediction horizon *n*	Number of states	Number of inputs	Number of iterations	CPU time (s)
MPC Delfland	24	900	97	4	1	1	10
scMPC DM Canal	1	60	61	6	1	5	1
scMPC IJmuiden	12	600	145	1	11	3	600
mmcMPC Delfland	24	900	97	10	1	1	120
MPC WM-Canal	2	240	31	37	7	1	10

	Remarks
MPC Delfland	Implemented as C-code in real-time on actual canal
scMPC DM Canal	Number of iteration is varying. Average number of iterations is 5
scMPC IJmuiden	Implemented as C-code in real-time on actual canal
mmcMPC Delfland	3 models in parallel
MPC WM-Canal	Implemented as C-code in real-time on actual canal

3.5 *Receding Horizon*

The receding or moving horizon principle is applied in Model Predictive Control. This means that the optimization is repeated every control time step and only the control actions calculated for the present time step are implemented. In the next control time step, the model is first updated from measurements and the most recent predictions are acquired. After that, the optimization is run. Using the most recent measurements of the water levels guarantees that the Model Predictive Controller acts like a feedback controller, bringing the measured deviations from setpoint back to zero. Using the most recent predictions of the disturbances such as rainfall-runoff, water level tide or offtake schedules, assures that accurate feedforward signals are used in the controller. In this way, the Model Predictive Controller has an up-to-date look into the future over the prediction horizon that recedes into the future with every time step the controlled water system makes.

3.6 *Conclusions on Model Predictive Control*

In this chapter all parts of a Model Predictive Control configuration are described. This configuration is setup specifically for water systems, as it captures the relevant dynamics and specifications of controlled water systems.
The most important conclusion is that the presented Model Predictive Controller can be applied to any type of open water system. Different systems, such as drainage systems or irrigation systems, only require a different internal model and another set of constraints. An additional conclusion is that the Model Predictive Controller can be tuned in a systematic way, by making use of physical characteristics of the water system.
The standard Model Predictive Control configuration can be implemented in a computer to control water systems in real-time.

4 Sequential configuration of Model Predictive Control

In the previous Chapter 3, the Model Predictive Control configuration applied in this research is described. This configuration utilizes a linear internal model and constant constraints. By using this simplification and applying quadratic programming, the optimization problem is convex. This implies that, if the solution is feasible, a global minimum with accompanying optimal control actions will be found.

Water systems of which the canal reaches can be modeled as delay time in series with a reservoir and the flows between these canal reaches are controlled by pumps, can be modeled as a linear internal model. In such case, the Integrator Delay model (Schuurmans J. (1997)) as described in Paragraph 2.2.2 can be applied. When the structures that discharge out of the water system and the structures that connect the canal reaches are variable speed pumps limited by a maximum pump capacity, constant constraints are valid. The standard Model Predictive Control configuration can be applied to control this simplified water system. When gates are applied as connecting structure or the canal reaches have dynamics that can not be modeled as a reservoir, non-linearities have to be taken into account in the model based controller. Another non-linearity that often occurs in water systems is the application of pumps that can only be turned off or on (Chow & Clarke (1994)).

In this chapter, an extended Model Predictive Controller is presented that can incorporate non-linear internal models and time-variant constraints. Instead of executing one optimization step, it utilizes sequel steps. With each step, the solution of the previous step is used to linearize the internal model and calculate the time-variant constraints. The configuration is referred to as sequential configuration of Model Predictive Control (scMPC).

4.1 Sequential linearization

The dynamics of water systems contain non-linearities. To be able to use standard Linear Algebra theory and software tools, it is convenient to convert these non-lineairities into lineair processes. The following major non-linearities of open water systems are identified:

- One-directional water flows and water levels in canal reaches are best described by the non-linear De Saint Venant equations (Chow (1959), Cunge et al. (1980), Stelling & Booij (1994)).;
- Gravity flow driven structures, such as undershot gates and overshot gates, are non-linear (Bos (1989), Brouwer (2004));
- Boundary conditions, such as tidal sea water levels, can make the boundary flows strongly time-variant.

In case trajectories of variables in time are available, the internal model can be linearized along these trajectories. By applying zero-order-hold sampling (Williamson (1991)) at each time step along the trajectories, the time-variant constraints can be calculated. When the linearized internal model and the time-variant constraints are applied in the optimization, the calculated control actions influence the trajectory of variables. In case large control time steps are applied, the linearization and constraints calculated at the previous step *k*, is not an accurate description of the changing dynamics over the total control time step at step *k*+1. A better approximation is to linearize at time step *k*+½. This can be implemented by weighing the new trajectory and the previous solution, with a weight factor ξ as presented in Formula 3.1. Using ξ=0.5, as proposed in the Crank Nicolson method (Stelling & Booij (1994)), gives the most accurate implementation. The optimization has to be repeated with the internal model linearized and constraints calculated along the weighted trajectories. The next trajectories have to be weighted again with the original trajectories. This loop has to be repeated, until the new trajectories do not change anymore. In the following example, the development of the trajectories is illustrated by an example of two reservoirs that are connected by a fixed submerged undershot gate. The initial trajectories are not known, so are considered constant at the initial water level.

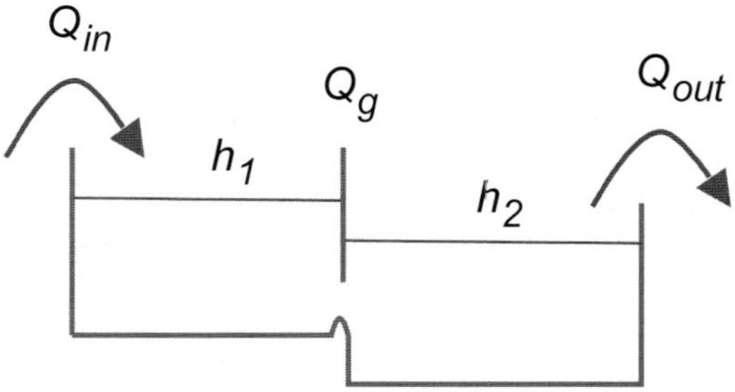

Figure 4.1 Two reservoirs connected by non-linear gate flow

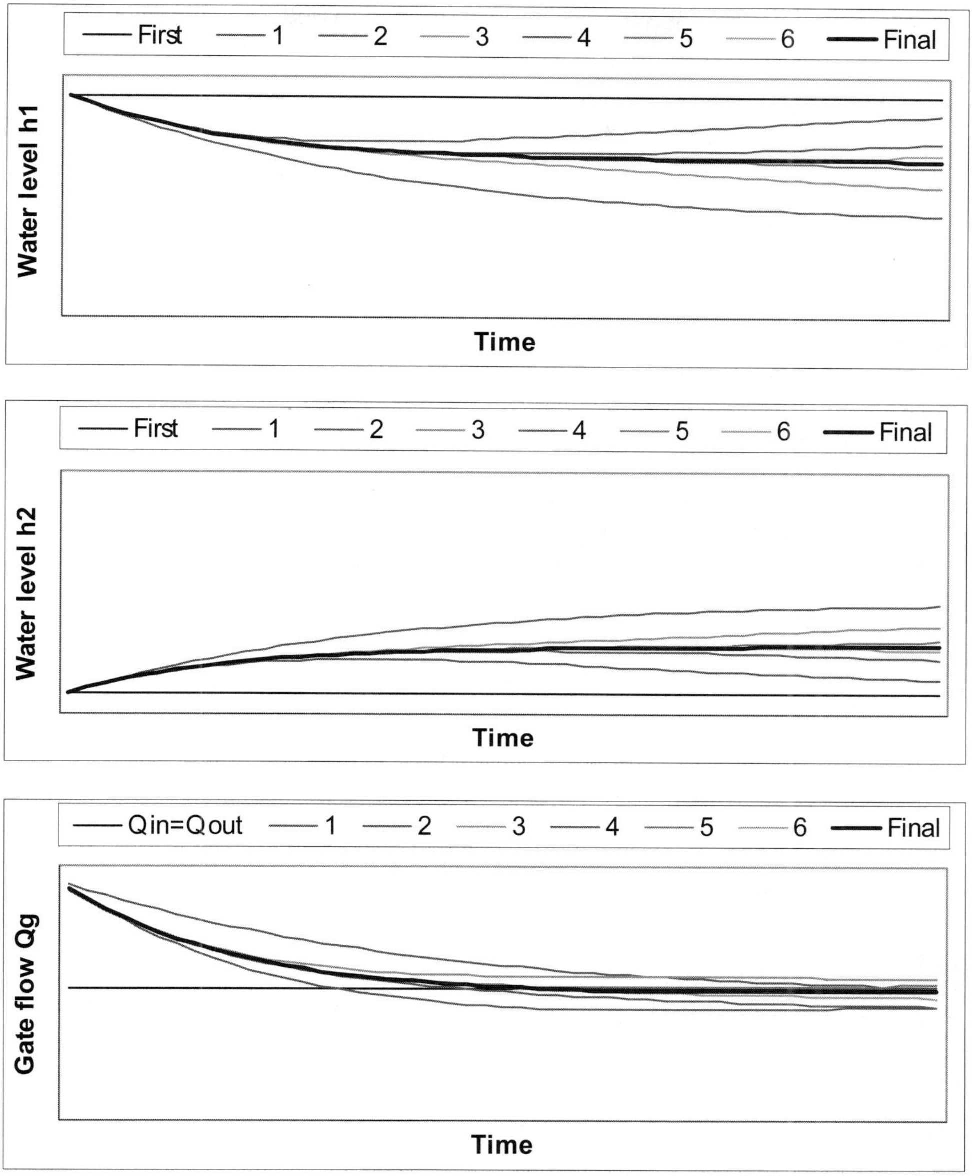

Figure 4.2 Trajectories of water levels and gate flow at each sequential step

In the results, the trajectories develop into the most accurate trajectory given the information that is available at the beginning of the simulation. At the end of the simulation the reservoirs are steady, as the final gate flow equals the constant in- and outflow. The iterative loop allows for implementation of non-linear processes in the standard Model Predictive Control configuration. The state space model as given in Formula 3.4 and the constraints presented in Formula 3.11, show this flexibility by the time-variant definition of the matrices and the constraints.

4.2 Sequential rounding

Water systems in low-land areas are often controlled by pumps that can either be turned off or on. The setup of the optimization as quadratic programming problem with continuous constraints as described in Chapter 3, does not allow for these discrete operational constraints. Other optimization techniques, such as constrained logic programming (Van Hentenryck, P. (1989)) and mixed integer programming (Tousain, R.L. (2002), Stork, M. (2005)), can apply search algorithms in the discrete solution area. Two disadvantages can be stipulated though, when these methods are implemented in real-time:

- The time that it takes for a discrete optimization algorithm to calculate a feasible solution, differs from moment to moment to a considerable extent. The time depends on the intelligence put into the search algorithm. Testing all combinations within a realistic time frame, is not possible. The application presented in Paragraph 6.3 has six off/on pumps. The number of combinations of these pumps being switch off or on over the prediction horizon of 72 steps, is 2^{432}.
- A result that is often observed with discrete solvers, is that a new solution is completely different compared to the previous solution. This is unacceptable for operators that use Model Predictive Control as a Decision Support System for manual control of a water system. In Paragraph 6.1 and Paragraph 6.3, these types of support systems are described. From interviews with operators, it showed that operational smoothness of the control actions in time is appreciated over the absolute lowest value of the objective function.

In this research, a sequential rounding algorithm is proposed. In a fixed number of sequential steps, the solutions are rounded to the discrete constraints. With every step, the margin around the discrete constraint in which a solution is rounded, is increased until all solutions are rounded. In Figure 4.3 an example of the off/on rounding algorithm of one pump over the prediction horizon is presented. Three iteration steps are applied in the example. In the first step, the rounding margin is $^{1}/_{3}{}^{rd}$ around the discrete constraint (off or on). In the second step the margin is $^{2}/_{3}{}^{rd}$. In the final step, the rounding margin around the discrete constraint is $^{3}/_{3}{}^{rd}$, so all solutions are rounded.

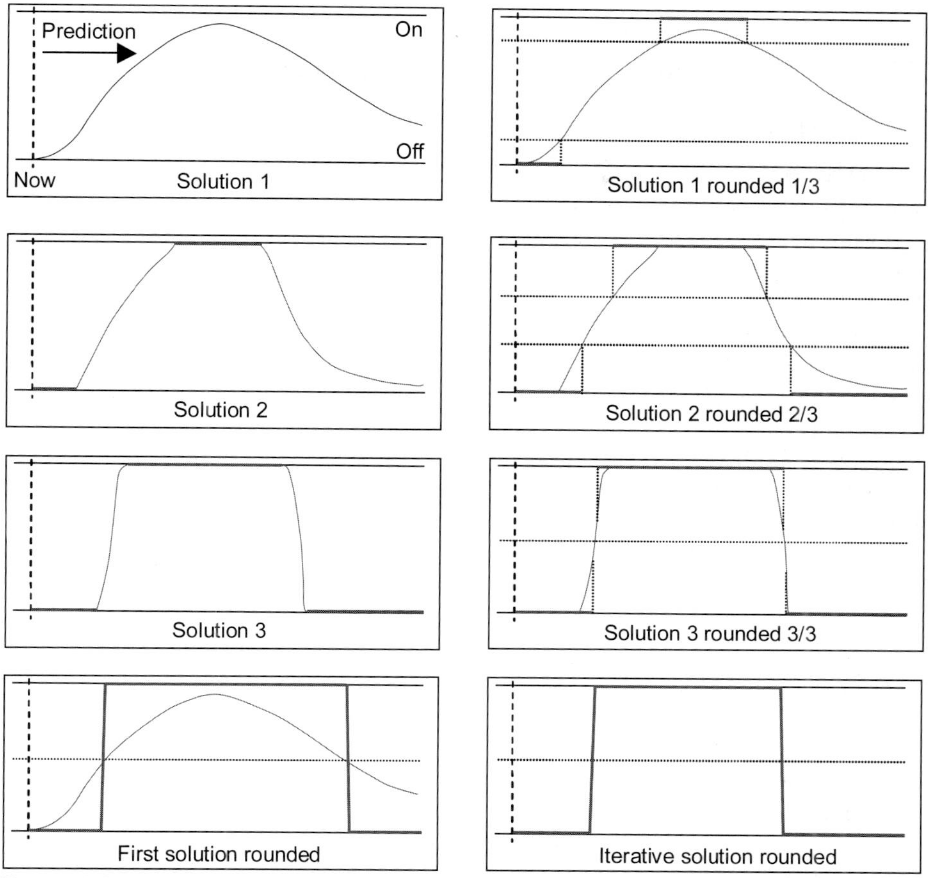

Figure 4.3 Step-wise rounding of pump flow with three iterations

An advantage of rounding in stages is the possibility for the optimization to compensate for the fixed solutions that result from rounding in a previous step. In this way, the optimization can keep the value of the objective function low and the solutions develop in a more discrete shaped trajectory. This can be seen by comparing the difference between the directly rounded solution and the iterative rounded solution in Figure 4.3. When the number of sequential steps is increased, the value of the objective function will approach the lowest possible objective function value given the discrete constraints.
Any type of discrete constraint can be incorporated in the sequential rounding algorithm. In the controlled water system as described in Paragraph 6.3, a constraint is implemented that requires pumps to remain in the off or on state for 30 minutes (Weissenbruch et al. (2004)). This constraint avoids the sequential on and off switching of the pumps to avoid wear and tear (see Figure 6.44). Another constraint that can be implemented is the requirement that a gate may only move

if the required change in gate height is larger than a certain minimum gate movement (Malda (2005)).

4.3 Conclusions on sequential configuration of Model Predictive Control

In this chapter, an extension to the standard Model Predictive Control configuration is presented that can deal with non-linearities in controlled water systems. The method is based on the repetition of the optimization at a given control time step, in a number of sequential steps. With every step in the Model Predictive Controller, the model is linearized and the time-variant constraints are determined again, depending on the newly computed trajectory of variables. The step-wise rounding towards a discrete solution, such as switching off/on pumps, can also be realized in the sequential configuration of Model Predictive Control. By using sequential configured Model Predictive Control, any type of non-linearity and constraint can be incorporated in the calculation of the optimal control actions. The consequence of the repetitions of the optimization in one control time step is the increase in computational time. The increase in computational time is approximately equal to the number of repetitions.

5 Multiple model configuration of Model Predictive Control

Model based control methods, such as Model Predictive Control, utilize internal models to predict future disturbances, inputs and states of the actual system. An inherent property of models is that their predictions are uncertain by definition. The model describes reality with simplified physical laws and is provided with uncertain boundary conditions and uncertain model parameters. The practical problem that results from the uncertainties in the internal model is that the actual behavior over a certain period is by definition never equal to the behavior that was predicted at the beginning of that period. Figure 5.1 present the rainfall-runoff flow into a drainage system at the beginning of a prediction horizon and the actual values of the runoff over that period. Due to accumulated errors, the prediction tends to drift away from the actual trajectory when the prediction moves further into the future.

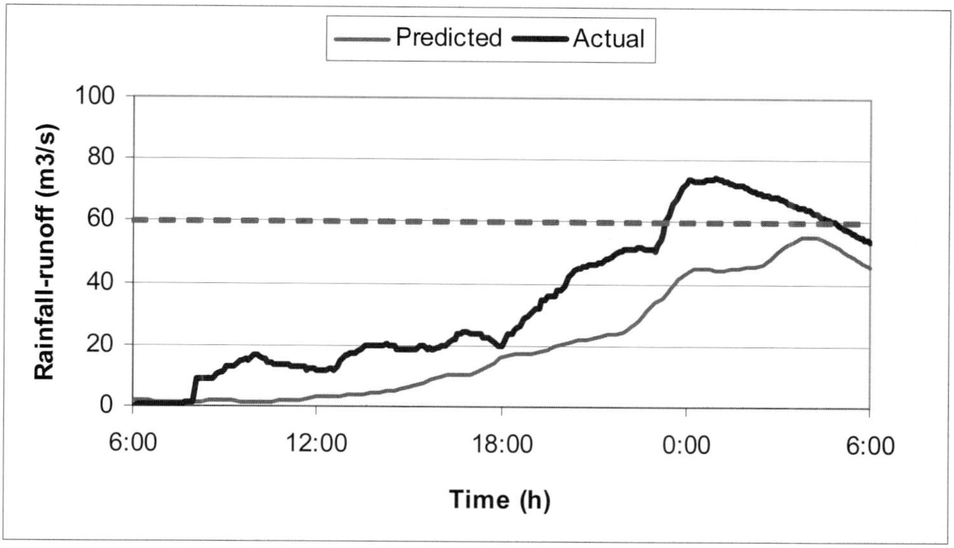

Figure 5.1 Predicted and actual rainfall-runoff

As the Model Predictive Controller uses a receding horizon, this decrease in prediction accuracy in time does not pose a problem. At every control step, new predictions are used that are more accurate for the same point in time compared to the predictions made at the previous control time steps. However, in case long prediction horizons and large control time steps are used, the inaccurate prediction further away in time will influence the present control actions in a negative way. In the example given in Figure 5.1, the predicted runoff of a storm event does not require the pumps to be turned on before the flow makes the water levels in the drainage system rise. The pump capacity of 60 m^3/s is sufficient to deal with this storm event. The actual rainfall-runoff does exceed the total pump capacity and requires premature lowering of the water levels in the drainage system. By doing so, extra storage is created in order to accommodate the runoff of the storm event. In case the incorrect control actions taken from inaccurate predictions can lead to high damage, it is important to obtain and to use an estimate of the extent of the incorrectness of the predictions in the calculation of the control actions. A good improvement to the given example would be to have an area of prediction solutions available of which the actual trajectory is part of (with a probability of e.g. 95%). Figure 5.2 depicts such a prediction area.

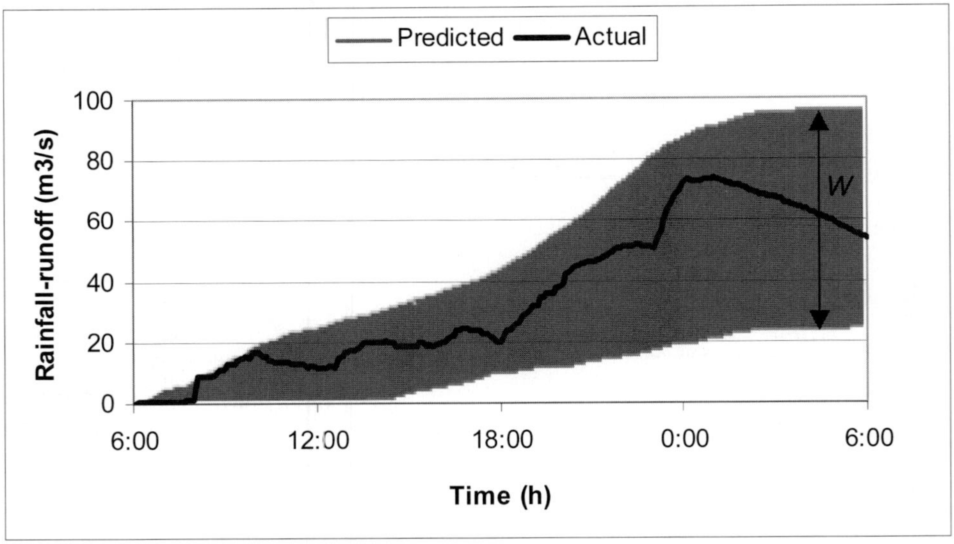

Figure 5.2 Actual and predicted area of rainfall-runoff

The width of this band can be used to set the relative importance of the result at the specific time step. This can be implemented by using the reciprocal of the width *W* of the uncertainty band at time step *k+i* based on knowledge available at control time step k in the objective function.

$$J = \sum_{i=0}^{n} \frac{1}{W(k+i \mid k)} \{\ldots\} \qquad \text{Formula 5.1}$$

In this research, a different, more stochastic approach is used. A Model Predictive Control configuration is presented that constructs the uncertainty area, converts this area into a minimum, average and maximum internal model, each with a certain fixed probability of occurrence and uses the different scenario results and their corresponding probabilities to calculate the control actions. The idea of replacing one model by using multiple models in parallel is based on Steinbuch (1989), Murray-Smith & Johansen (1997), Overloop et al. (2005b).

5.1 Uncertainties in water system models

The uncertainties of the internal model used in Model Predictive Control are caused by three types of errors:

- Errors in the model structure. Usually, the model structure is simplified to a considerable extent. This simplification is unavoidable in order to achieve a certain computational speed required to calculate the control actions in real-time;
- Errors in the boundary conditions. The boundary conditions are usually not exactly known in advance, but nevertheless introduced in models are given values. Improvements can be made if boundary conditions are treated as stochastic variables and used to create an ensemble of predictions by running the model numerous times with different input values. For example (Kruizinga, S. (2002)) uses fifty runs to make ten day precipitation forecasts;
- Errors in the parameters of the model. Model parameters are, like boundary conditions, stochastic variables, but usually considered fixed. The mean values of the parameters are found by measurement of the dimensions of the (sub-)system, calibrated from measurement series or estimated. The parameters that are found from direct measurements of a physical property have a low uncertainty. An example of such a parameter in a drainage system is the catchment's surface of unpaved area. In canal reaches, the cross-sectional profile is a parameter that can be measured accurately. Parameters found from calibration on measurement series have a much higher uncertainty e.g. the storage capacity in the ground of the drainage system or the bed friction value in a canal reach. This can be due to inaccurate measurement series or from the simplification that the parameters of more processes are combined into one parameter that is used in the internal model (lumped parameter).

As the model structure is a given fact and the boundary conditions are available from an outside source, only the third type of errors, the parameter uncertainties, can be estimated and used to improve the water management.
All of the errors cause the actual parameter value to differ to a certain extent from the selected mean parameter value. A general way to describe the uncertainty of parameters is by using a Normal distribution (Gehrels (2003)). This probabilistic distribution is determined by the mean value μ and the standard deviation σ. In Formula 5.2 the Normal distribution of parameter x is given.

$$P(x) = \frac{1}{\sigma\sqrt{2\cdot\pi}} e^{\frac{-(x-\mu)^2}{2\cdot\sigma^2}}$$ Formula 5.2

Figure 5.3 presents the Normal distribution graph for μ=0 and σ=1.

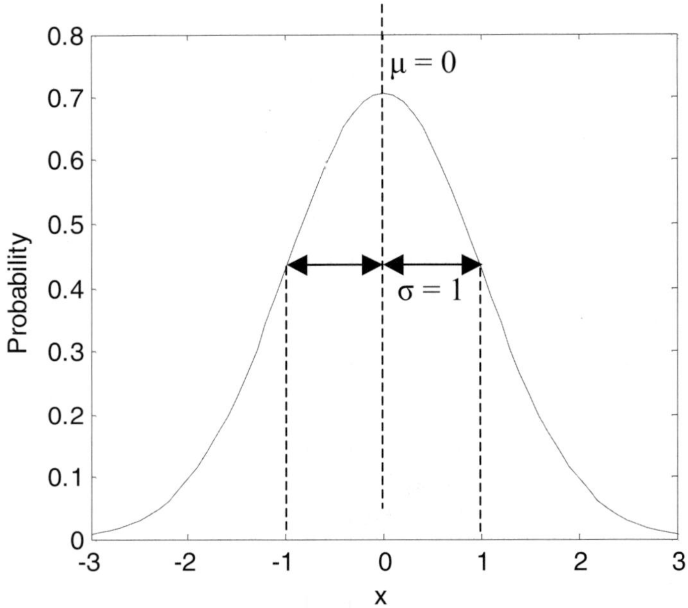

Figure 5.3 Normal distribution graph of parameter x (μ=0, σ=1)

In the given example, the catchment's area has a low standard deviation, relative to the mean value, while the storage capacity in the ground has a large relative standard deviation.
In this research, the focus is on the model used in the feedforward part of the controller. This model describes the disturbance processes. This choice is made based on the fact that especially rainfall-runoff processes that disturb drainage systems are complex (non-linear) and uncertain to a considerable extent.
A generally accepted way to estimate the uncertainty of a model is to use a Monte Carlo analysis (Sobel' (1994), Manno (1999)). Here, the calculation of the prediction over the same period is repeated a number of times. In every calculation, all parameters are changed. The parameter values are randomly picked from their uncertainty distribution. The final result of the Monte Carlo analysis is a spectrum of solutions, referred to as confidence interval. These solutions can be considered as the uncertainty area that is presented in Figure 5.2. Other methods that try to calculate the uncertainty of the prediction directly from the uncertainties of the parameters are hard to apply to water systems. This

is due to the non-linear behavior of water systems, especially of the rainfall-runoff processes (Savenije H.H.G. (2001)).
The ultimate usage of the calculated confidence limits is to use all solutions in the Model Predictive Controller. Due to the limitation of the computational speed, this is not an option. Instead, a minimum, average and maximum model is derived from the solutions. In this research, the three models are derived from the assumption that the solutions can be sorted in a certain way and a group of sorted solutions can be assembled by taking the average trajectory of the grouped solution.
In the feedforward part of all MPC applications described in Chapter 6, the disturbances constitute all flows, either in or out of the water system. These flows cause the water levels to drift away from their setpoint. The integrated flow, i.e. the volume, over the prediction horizon determines the change in water level. Consequently, sorting the solution on volume is the most logical choice. A different option is to also pick the parameters of the process model randomly from a Normal distribution, apply feedback and feedforward and sort on the resulting water level rise.
An additional assumption that is made in the derivation of the three models is that the minimum and maximum model are taken from the average of the lowest and highest 10% of the sorted solutions, while the average model is taken from the average of the middle 80% of the sorted solutions. This is an assumption that is more often applied in management of water systems, but it is merely an assumption. This assumption before the actual usage of the models in a stochastic calculation weakens the correctness of the final result. A different selection of models will lead to (slightly) different calculated control actions.

5.2 Controlling uncertain water systems

Model Predictive Control uses the internal model of the water system to calculate optimal control actions over the prediction horizon at each control time step. For uncertain water systems there are now three models, a minimum, average and maximum model, available with different probabilities of occurrence and different trajectories of the water levels. Figure 5.4 presents the transformation from one model into three probabilistic models. An important given fact is that a controller only calculates one trajectory of control actions. In Figure 5.5 the results of the three models are presented controlled by one trajectory of control actions.

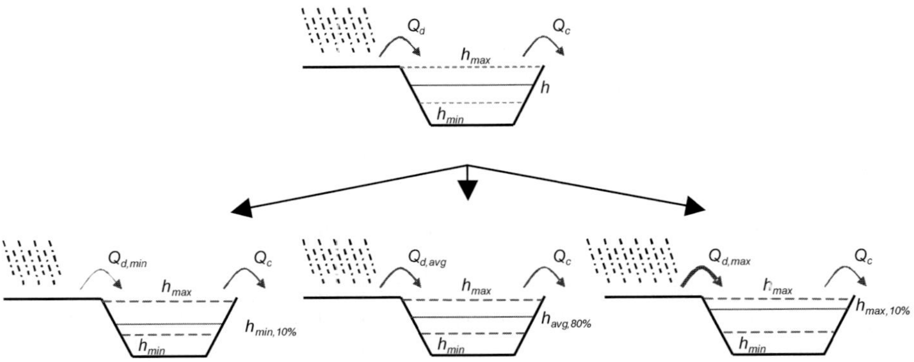

Figure 5.4 Internal model transformed into three models with different probability

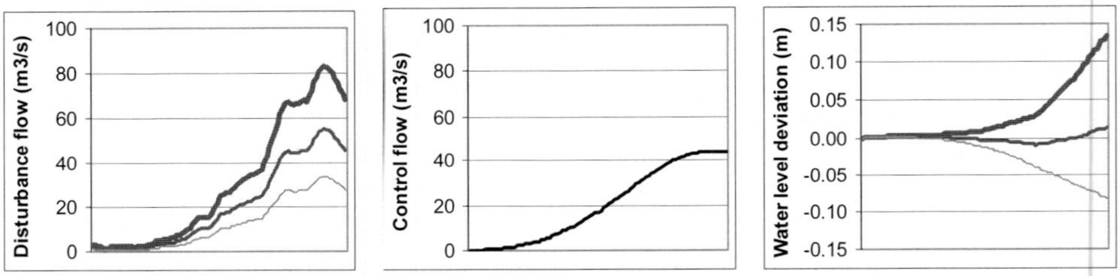

Figure 5.5 Three models controlled by one trajectory of control actions

The deviation of the water level from setpoint is the main controlled variable in water systems, so it is part of the objective function used in MPC. The objective function can be considered to be built up from components that describe the cost. Note that in other research areas, the objective function is often named Cost function. For the given example, there now is a minimum model with high cost and low probability, an average model with low cost and high probability and a maximum scenario with high cost and low probability. A way to combine probability and cost is to use the variable risk (Holton (2004)). The most widely accepted definition of risk is that risk equals probability times cost (Penning-Rowsell & Chatterton (1977)). This allows for incorporation of all three models in one objective function. The components in the objective function that are linked with the water level deviation, are extended to three water level deviations and multiplied by their respective probability. Formula 5.3 describing (a part of) a general water management objective function, changes into Formula 5.4 that captures the risk-approach.

$$J=\sum_{i=0}^{n}\left\{e(k+i\mid k)^{T}\cdot Q\cdot e(k+i\mid k)+\ldots\right\}+\ldots \qquad \text{Formula 5.3}$$

$$J=\sum_{i=0}^{n}\left\{Q\begin{pmatrix} e_{avg}(k+i\mid k)^T\cdot P_{avg}\cdot e_{avg}(k+i\mid k)+\\ e_{\min}(k+i\mid k)^T\cdot P_{\min}\cdot e_{\min}(k+i\mid k)+\\ e_{\max}(k+i\mid k)^T\cdot P_{\max}\cdot e_{\max}(k+i\mid k)\end{pmatrix}+\ldots\right\}+\ldots \quad \text{Formula 5.4}$$

where P represents the probability of occurrence of a scenario. As multiple models are applied in parallel, this MPC configuration can be referred to as multiple model configuration of Model Predictive Control (mmcMPC) or Multiple Model Predictive Control. In Rao et al. (1999) and Gonzalez Santos et al. (2000) multiple models are used in a model based controller. However, these applications do not use the risk-approach as used in this research.

5.3 Application of multiple model configuration of Model Predictive Control

By using a stochastic objective function, the control actions are mainly based on the average model, as this scenario has the highest probability. Only when the minimum or maximum scenarios result in a very high cost, the control actions will be based more on this costly scenario. This is an improvement over the regular MPC configuration.
An additional improvement can be implemented by the application of output constraints on the resulting water levels of one of the models. In a drainage system, there can be a maximum water level constraint on the water level of the maximum model in order to avoid flooding under all circumstances. In Figure 5.6 the result of the application of output constraints on the minimum and maximum models are presented. Compared to the multiple model configuration of MPC configuration of which the results are given in Figure 5.5, the water levels are lowered more to avoid exceeding of the maximum allowed water level by the maximum model.

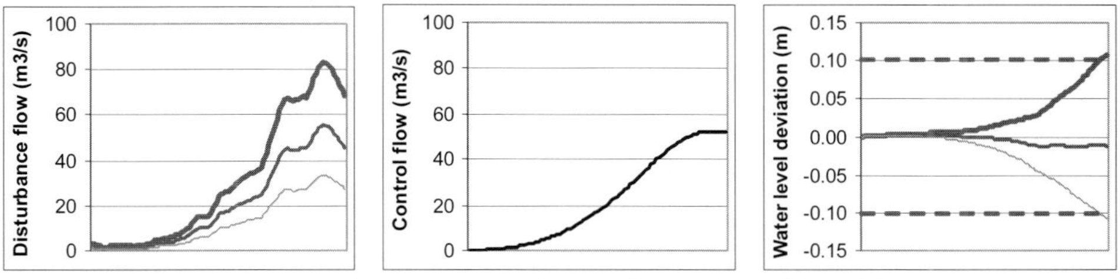

Figure 5.6 Results of multiple model configuration of MPC with water level constraints on the minimum and maximum model

The extra functionalities that multiple model configuration of MPC offers result in an increase in the dimensionality of the optimization problem. For example, the size of the system matrix of the internal model increases with a factor 3^2. In the application described in Paragraph 6.4 this results in an increase in computational time of a factor 12 (see Table 3.1).

5.4 Conclusions on multiple model configuration of Model Predictive Control

Multiple model configuration of Model Predictive Control is a Model Predictive Controller that can incorporate uncertainties of water system models in the calculation of the optimal control actions by using the risk-approach in the objective function. It does not differ from the way standard MPC is setup. It only utilizes minimum, average and maximum models in parallel in the internal model, with a probability of 10%, 80% and 10%, respectively. Although the computational time increases considerably, the actual implementation is feasible on fast computers.
The transformation of the uncertainty and the objective (cost) function into a risk objective function, allows for a deterministic solution to a stochastic control problem.

6 Applications of Model Predictive Control

In the previous chapters, the theory of controlled water systems and the various Model Predictive Control configurations are described. The question remains, what control method has to be selected given the objective of the water management and the characteristics and constraints of the water system? To support the selection of the appropriate control method, a procedure is developed, that can be used as an expert system. The procedure is the result of the knowledge gained in this research. The foundation of the procedure is the thesis that the control method has to be as simple as the objective, characteristics and constraints allow it to be. In case local feedback control satisfies the specifications, this control method has to be selected over more advanced control methods, especially when application of communication lines can be avoided (prone to failure). Only when constraints make it impossible to satisfy the specification or when the specifications become more stringent, Model Predictive Control methods come into play. The control method selection procedure as developed in this research is presented in Figure 7.1 as recommendation expert-system for future work in the field of control of water systems.
In this chapter, five different Model Predictive Control configurations are applied to different types of water systems. Below, the reasoning behind the selection of the types of water systems is given in order to substantiate that Model Predictive Control improves the water management of different types of water systems. Also, the selection of the control method, related to the objective, characteristics and constraint that apply to the controlled water system, is argued (Figure 7.1).

The five Model Predictive Control applications are:

- Advisory system for drainage system Delfland. This drainage system is the fastest responding large drainage system in The Netherlands. It has recently suffered from extreme storm events in 1998, 1999, 2001 and 2002, resulting in high economic damage. The storage canals are controlled only by pump stations. The total pump capacity is limited and insufficient for heavy storm events. The constraints on the pump stations require the application of Model Predictive Control;
- Local control of irrigation canal reach Delta Mendota. This application is presented to demonstrate that MPC can also function as a local controller and has a higher performance than the less advanced control methods feedback and feedforward. The water flows and water levels along the canal reach are estimated by using a De Saint Venant internal model. This requires linearization of the De Saint Venant equations in a sequential loop (sequential configuration of MPC);
- Off/on control of pump station IJmuiden. The pump station IJmuiden is at present the largest pumping installation in Europe. Every day, it uses a considerable amount of energy. An advanced control method such as Model Predictive Control is required to minimize safety from inundation against minimum energy costs. The pumps can be either turned off or on.

These constraints require a rounding procedure that, in this research, is implemented in a sequential loop (sequential configuration of MPC);

- Stochastic control of drainage system Delfland. The rainfall run-off processes that cause the water levels in the storage canals of Delfland to rise, cannot be modeled accurately. The parameters in this disturbance model are uncertain to a considerable extent. For that reason, the standard Model Predictive Controller is extended with multiple models (multiple model configuration of MPC) for a minimum, average and maximum scenario with different probabilities;
- Centralized control of irrigation canal W-M. This application is presented to proof that Model Predictive Control can function in real-time on a total canal and that both the standardized MPC configuration and the general tuning procedure are valid.

In the following paragraphs, each water system and MPC application is described in detail. Per application, the objective of the water management, the characteristics and constraints of the water system are described. Subsequently, the results are presented and discussed.

6.1 Advisory system for drainage system Delfland

The main function of the large canals in the district of the water board of Delfland in the Western part of The Netherlands is to gather excessive water from polders and to discharge this water to the surrounding rivers and the sea. Figure 6.1 shows the location of water board of Delfland on a map of The Netherlands. To discharge water, the water managers have six large pump stations at their disposal. Figure 6.2 shows a schematic overview of the district. The lines represent the storage canals, the numbers outside of the area the six pump stations and the numbers inside of the area the five measurement locations (Schuurmans, W. et al. (2003a), Schuurmans, W. et al. (2003b)).

Figure 6.1 Water board Delfland on map of The Netherlands

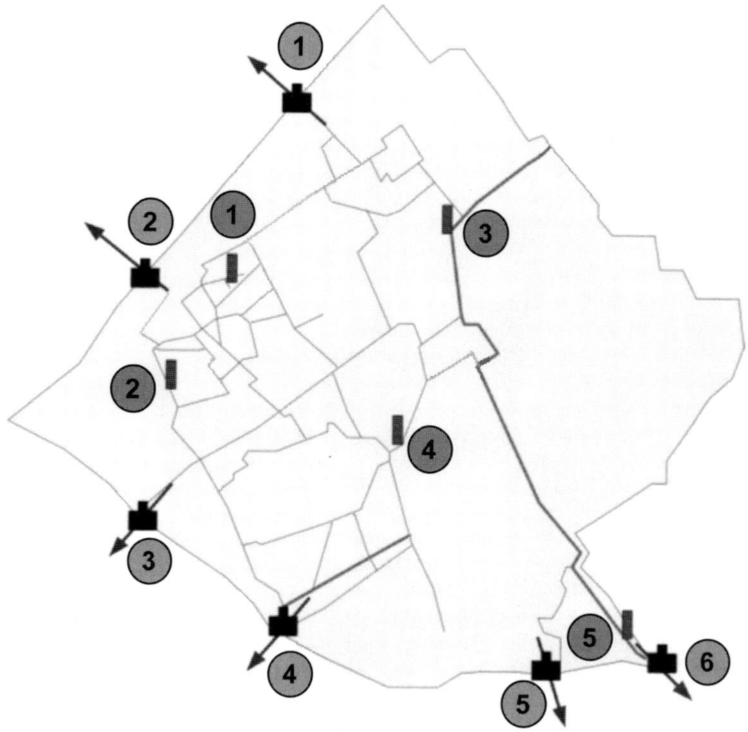

Figure 6.2 Schematic overview of drainage system Delfland

The excessive water is the result of runoff from precipitation in the area. The hydrological run-off water flows via various paths (Savenije,H.H.G. (2001)). Some precipitation falls directly on the canals (fastest component). Some precipitation falls on paved areas and is discharged to the canals through rain water sewer systems (fast component). Some precipitation infiltrates in unpaved areas, causing the ground water table to rise and flow out of the soil to the ditches in the unpaved area. If the ground water levels become high or the precipitation is higher than the infiltration capacity, surface run-off into the ditches occurs. Most of the area lies lower than the drainage canals. The water in the ditches of these areas is pumped into the storage canals (slow component). Other areas have a higher elevation than the drainage canals and flow through gravity driven structures into the drainage canals (fast component). Figure 6.3 presents the hydrological runoff. Figure 6.5 shows a typical total inflow into the storage canals after the storm event at March 13th 2005 as given in Figure 6.4 (Weijs (2004)). March 13th 2005 is selected as the test period in the first simulations that are presented in Paragraph 6.1.4. The reason for selecting this period is that all measurement data and weather forecasts are available for testing. An additional reason is that it is a typical case in which the amount of forecasted and actual precipitation is approximately the same, but the intensity of the actual precipitation at a peak moment is much higher. This situation often occurs.

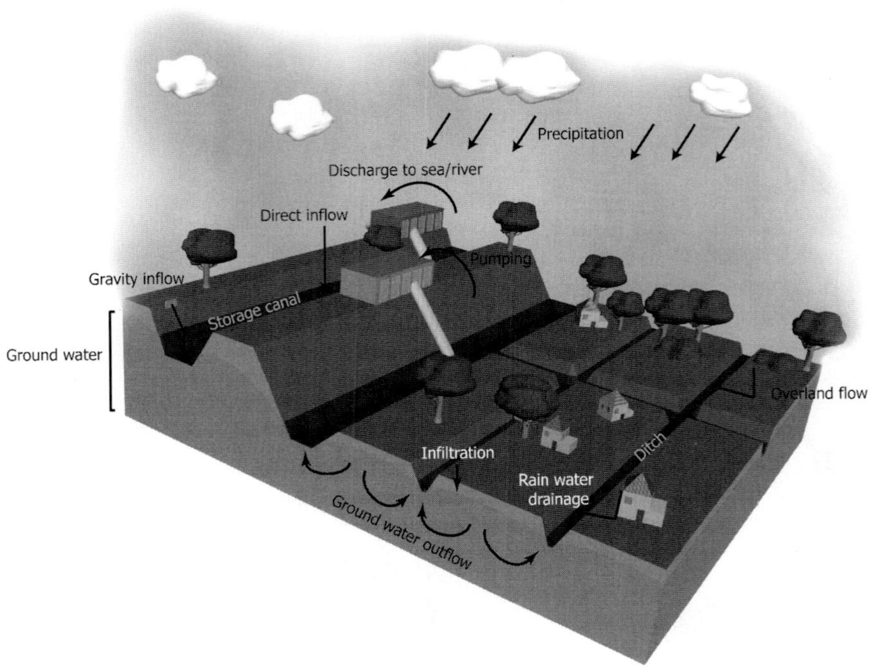

Figure 6.3 Hydrological rainfall-runoff

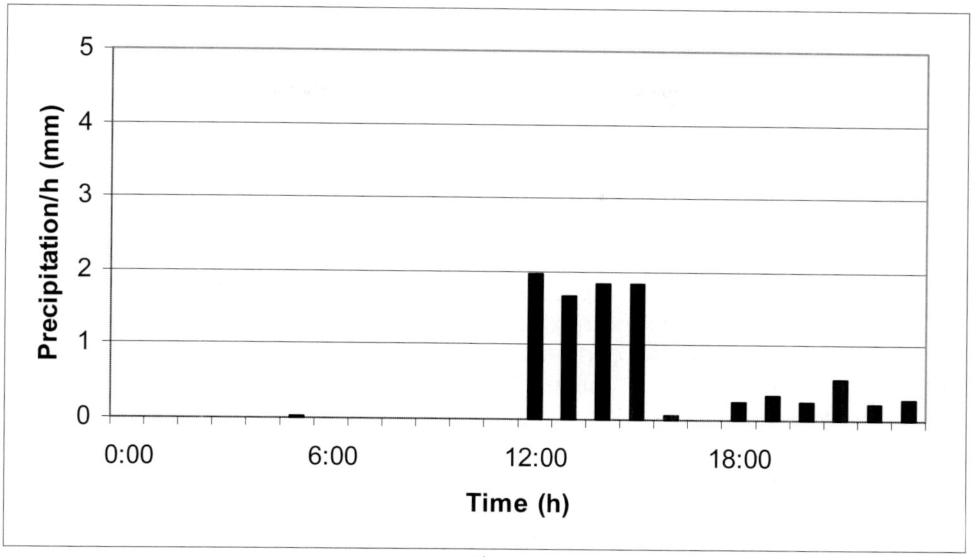

Figure 6.4 Precipitation of March 13th 2005

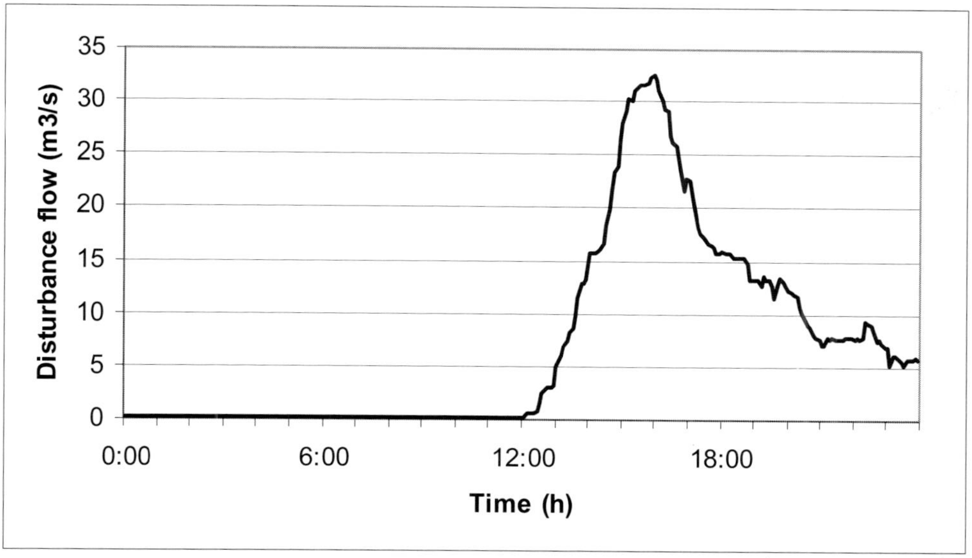

Figure 6.5 Rainfall runoff flow into storage canals (March 13th 2005)

6.1.1 Objective of control on drainage system Delfland

The objective of the control on the storage canals is to keep the water levels in the canals within a range around setpoint. This range is given by a maximum

allowed water level and minimum allowed water level. The maximum allowed water is set to avoid (risk of) overtopping of dikes, preventing possible inundation of the surrounding land. The minimum water level is set to avoid (risk of) de-stabilizing dikes and damaging houseboats. During regular operation, the water level is controlled close to setpoint. A secondary objective is to change the pump flows as few times as possible (of course without violating any water level constraints).
The water level is measured at five locations and transmitted to the central operating room. Here, a representative water level (RWL) is automatically computed by weighing these measurements. A short-term precipitation forecast is available for every hour for the next 24 hours. Also, from this central location the various pumps in the district can be remotely controlled. The operator in charge is responsible for these control actions. Since 2003, a control system based on the Model Predictive Controller described below, is operational in the central operating room of the water board of Delfland. It functions as a Decision Support System rather than a fully automated control system.

6.1.2 Characteristics of drainage system Delfland

The drainage system of Delfland is the fastest reacting drainage system in The Netherlands. The main reason is the high density of green houses. Within 30 minutes after a storm event starts, the water level in the storage canals starts to rise. The water managers base their control action on forecasts of 24 hours. The trade-off between computational time over this prediction horizon and the level of detail of the control actions results in a control time step T_c of 15 minutes.
The storage canals are wide and consequently have a high capacity. As many canals are linked together in a network and can easily communicate water flows, the canal system can be considered as one large reservoir. Figure 6.6 shows a representation of the mass balance of the water system.

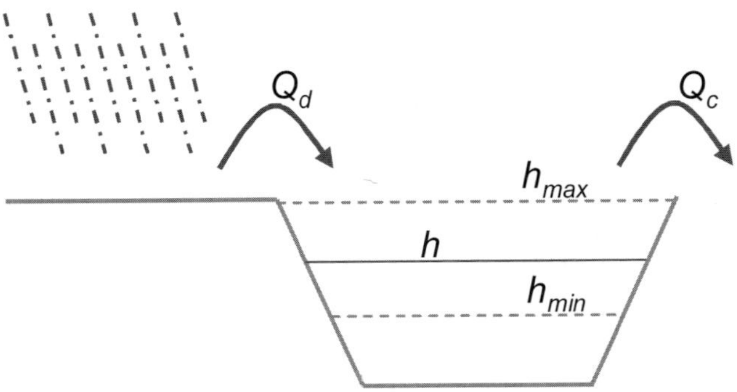

Figure 6.6 Mass balance of storage canals Delfland

Q_d represents the inflow into the storage canals. This flow is considered as a disturbance on the water system. Q_c represents the total control flow that the six pump stations discharge out of the storage canals, h the representative water level and h_{min} and h_{max} the minimum and maximum allowed water levels. The delay time T_d between a change of the pump flow and the resulting change in the representative water level is approximately 15 minutes, which is equal to one control time step. The physical explanation for this delay time can be found in the simplified modeling of the water system as a reservoir with one representative water level. The representative water level is the weighted result of five water level measurements at distributed locations. This filtering introduces extra lag time. In Formula 6.1 the linear discrete model is formalized according to the theory as given in Paragraph 3.1:

$$h(k+1) = h(k) - \frac{Q_c(k-k_d)\cdot T_c}{A_s} + \frac{Q_d(k)\cdot T_c}{A_s} \qquad \text{Formula 6.1}$$

where k represents the discrete time step index, T_c the control time step (s), k_d the number of delay steps between control action and water level rise and A_s the storage area of all canals at setpoint (m^2).
The deviation e between water level h and the fixed setpoint h_{ref} is defined as follows:

$$e(k) = h(k) - h_{ref} \qquad \text{Formula 6.2}$$

The change in control flow is used as a function of the change in water level and the water level itself, comparable to a Proportional Integral controller. In this way, zero static deviation can be achieved. The change in control flow and change in water level are defined as:

$$\Delta Q_c(k) = Q_c(k) - Q_c(k-1) \qquad \text{Formula 6.3}$$

$$\Delta e(k) = e(k) - e(k-1) \qquad \text{Formula 6.4}$$

Formula 6.5 shows the general state space representation of the controlled water system:

$$\begin{bmatrix} e(k+1) \\ \Delta e(k+1) \\ \Delta Q_c(k) \end{bmatrix} =$$

$$\begin{bmatrix} 1 & 1 & -\frac{T_c}{A_s} \\ 0 & 1 & -\frac{T_c}{A_s} \\ 0 & 0 & 0 \end{bmatrix} \cdot \begin{bmatrix} e(k) \\ \Delta e(k) \\ \Delta Q_c(k-1) \end{bmatrix} + \begin{bmatrix} 0 \\ 0 \\ 1 \end{bmatrix} \cdot [\Delta Q_c(k)] + \begin{bmatrix} \frac{T_c}{A_s} \\ \frac{T_c}{A_s} \\ 0 \end{bmatrix} \cdot [\Delta Q_d(k)]$$

Formula 6.5

6.1.3 Constraints on drainage system Delfland

The following constraints apply to the controlled water system:

- A minimum control flow $Q_{c,min}$ of zero. No water is pumped into the system during drainage control;
- A maximum control flow $Q_{c,max}$ equal to the maximum capacity of all six pump stations together;
- A minimum change in control flow $\Delta Q_{c,min}$ equal to a certain flow step change. In case the required flow change is smaller than this value, the control action is postponed;
- A minimum water level h_{min} equal to the minimum allowed water level;
- A maximum water level h_{max} equal to the maximum allowed water level.

The minimum and maximum control flows are set as hard constraints on the input Q_c. The minimum change in control flow is not incorporated in the optimization as such a constraint requires sequential configuration of Model Predictive Control. Instead, the computed control flow is rounded afterwards to the nearest multiple of the minimum change in control flow. In this rounding a dead band is used to avoid sequential switching between two solutions. The minimum and maximum water levels are set as soft state constraints on the deviation *e* outside of this range (Boom & Backx (2001), Hovland (2004)). The state space model including the constraints becomes:

$$\begin{bmatrix} e(k+1) \\ \Delta e(k+1) \\ e^*(k) \\ Q_c(k) \end{bmatrix} =$$

$$\begin{bmatrix} 1 & 1 & 0 & -\frac{T_c}{A_s} \\ 0 & 1 & 0 & -\frac{T_c}{A_s} \\ 1 & 1 & 0 & -\frac{T_c}{A_s} \\ 0 & 0 & 0 & 0 \end{bmatrix} \cdot \begin{bmatrix} e(k) \\ \Delta e(k) \\ e^*(k-1) \\ Q_c(k-1) \end{bmatrix} + \begin{bmatrix} 0 & 0 \\ 0 & 0 \\ 0 & -1 \\ 1 & 0 \end{bmatrix} \cdot \begin{bmatrix} \Delta Q_c(k) \\ u^*(k) \end{bmatrix} + \begin{bmatrix} \frac{T_c}{A_s} \\ \frac{T_c}{A_s} \\ \frac{T_c}{A_s} \\ 0 \end{bmatrix} \cdot \left[\Delta Q_d(k) \right]$$

$$Q_c(k) \geq Q_{c,\min}(k)$$
$$Q_c(k) \leq Q_{c,\max}(k)$$
$$u^*(k) \geq h_{\min}(k) - h_{ref}(k)$$
$$u^*(k) \leq h_{\max}(k) - h_{ref}(k)$$

Formula 6.6

where e^* represents the water level outside of the allowed range around setpoint and u^* the virtual signal that is subtracted from the water level deviation to make e^* either zero or a value equal to the exceeding of the range around setpoint (Boom & Backx (2001)). The parameters of the controlled water system are given in Table 6.1.

Table 6.1 Parameters of controlled water system

Parameter	Value	Unit
T_c	900	s
A_s	7.30e6	m^2
T_d	900	s
h_{ref}	-0.40	mMSL
h_{min}	-0.55	mMSL
h_{max}	-0.30	mMSL
Q_{min}	0.0	m^3/s
Q_{max}	60.0	m^3/s
ΔQ_{min}	2.0	m^3/s

6.1.4 Results of control on drainage system Delfland

To demonstrate the potential that Model Predictive Control has to control water systems, it is compared to other less advanced control methods namely feedback control and the combination of feedback and feedforward control. The comparison is made in Paragraph 6.1.5.
All control methods are tested on a calibrated model (Sobek (2000)) of the runoff processes and hydro-dynamics of the canals. The different controllers are programmed in Matlab (MathWorks (1992)). The precipitation and the resulting inflow of the March 2005 storm event into the storage canals are shown in the previous given Figures 6.4 and Figure 6.5, respectively.

First, Feedback control is applied using a Proportional Integral controller tuned with tuning rules for delay time dominated water systems (Schuurmans, J. (1997)):

$$\Delta Q_c(k) = K_p \cdot \Delta e(k) + K_i \cdot e(k) \qquad \text{Formula 6.7}$$

where K_p represents the proportional gain (=1106.1) and K_i the integral gain (=75.4).

Figure 6.7 shows the flow of the six pump stations that discharge the water out of the storage canals. Figure 6.8 shows the representative water level and the water levels from which the RWL is computed.

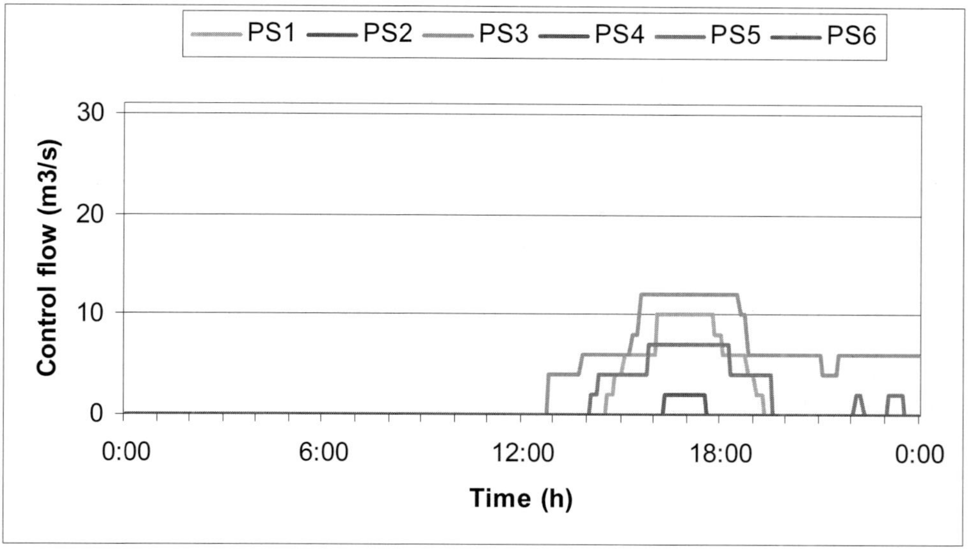

Figure 6.7 Discharge of pump stations when feedback control is applied (March 13th 2005)

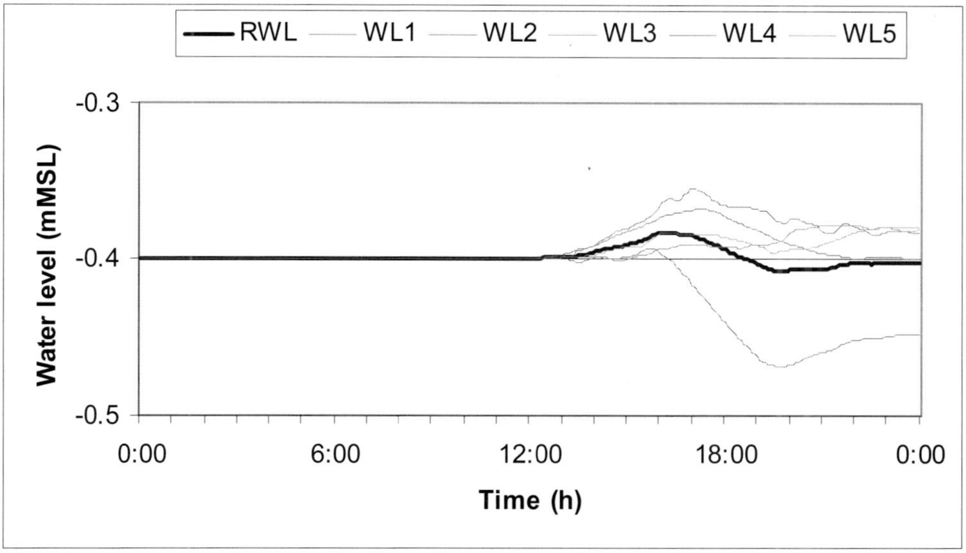

Figure 6.8 Representative water level and distributed water levels when feedback control is applied (March 13th 2005)

Next, the feedback controller is extended with feedforward control. The feedforward controller tries to exactly cancel the change in the water level caused by the disturbance inflow. The transfer function between control flow Q_c and disturbance flow Q_d has an amplitude of 1, but has a phase lag of plus one time step. This non-causality requires the use of a prediction of the disturbance flow ΔQ_d at time step k+1:

$$\Delta Q_c(k) = K_p \cdot \Delta e(k) + K_i \cdot e(k) + \Delta Q_d(k+1) \qquad \text{Formula 6.8}$$

To construct the predicted disturbance flow, the forecast of the precipitation for the next 24 hours is used. Figure 6.9 shows the measured precipitation as given in Figure 6.4 together with the forecast of the storm event at 0 hours, 6 hours, 12 hours and 18 hours, respectively. The closer the prediction is to the actual event, the more accurate the prediction is supposed to be. The forecast of the precipitation is used in the disturbance model to compute a prediction of the runoff flow into the storage canals. Figure 6.10 shows the actual inflow as used in the tests together with the predictions of the inflow at 0 hours, 6 hours, 12 hours and 18 hours, respectively as used by the controller.

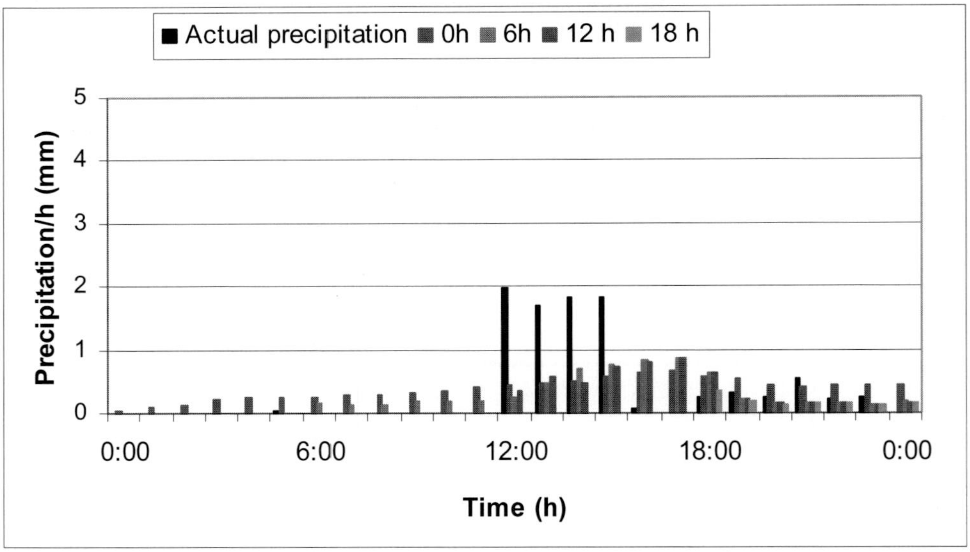

Figure 6.9 Actual and forecasted precipitation (March 13th 2005)

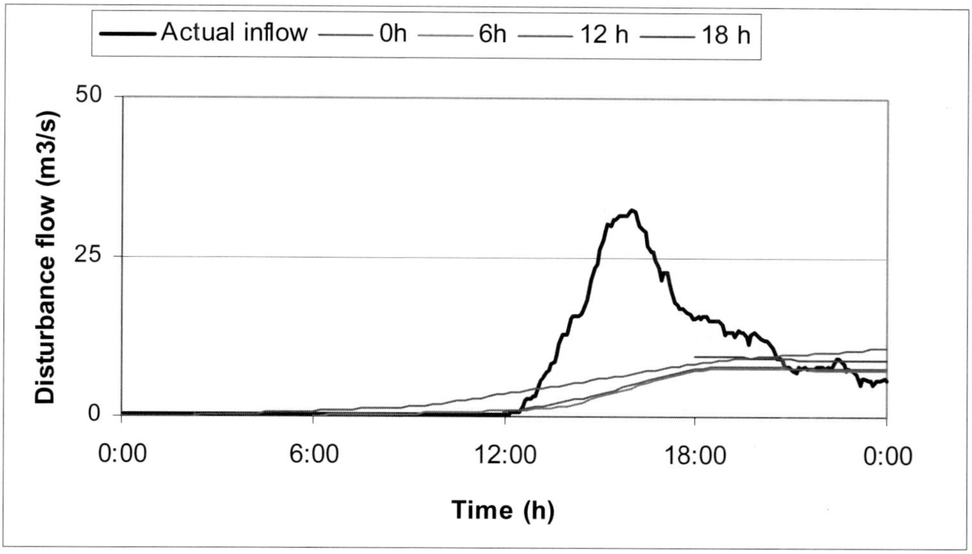

Figure 6.10 Actual and predicted disturbance (runoff) flow (March 13th 2005)

Figure 6.11 and Figure 6.12 show the discharge of the pump stations and the resulting water levels when feedback in combination with feedforward is applied.

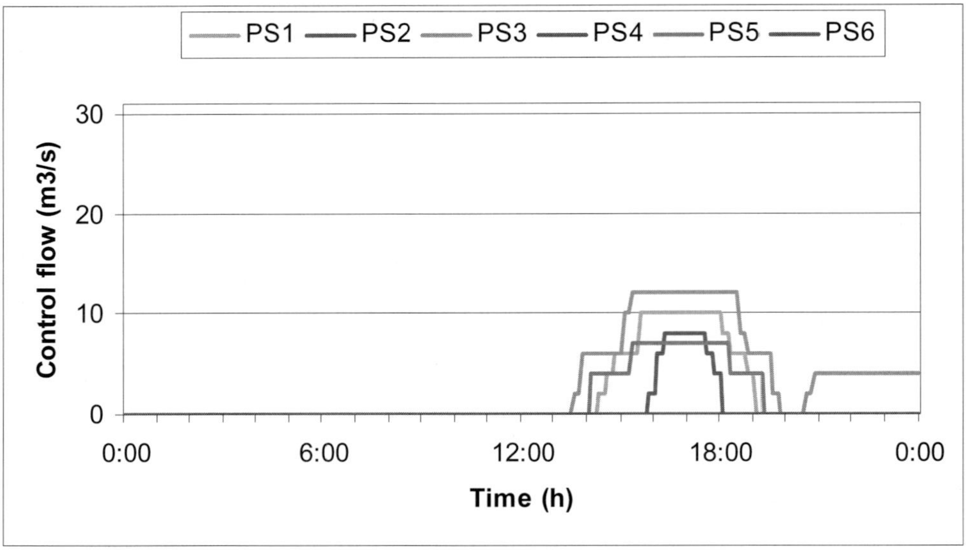

Figure 6.11 Discharge of pump stations when feedback and feedforward control is applied (March 13th 2005)

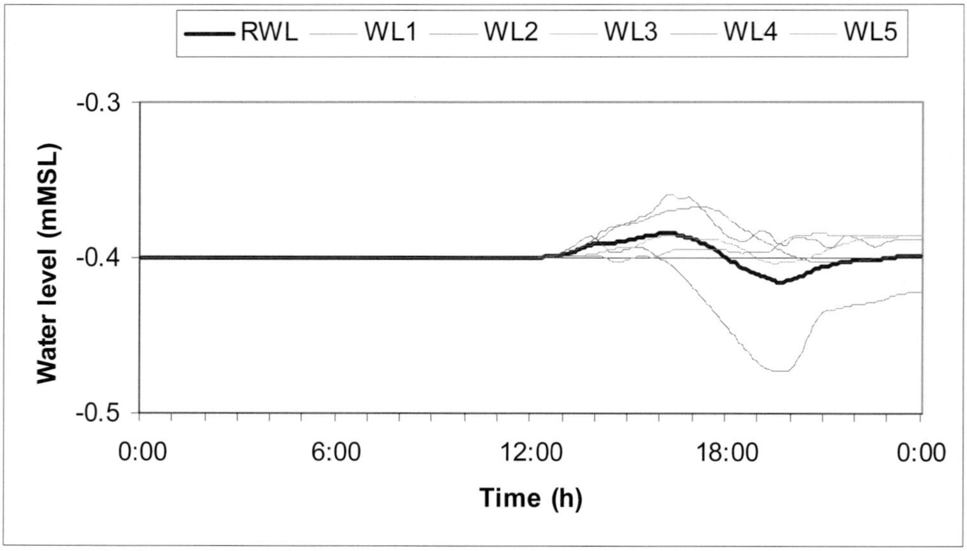

Figure 6.12 Representative water level and distributed water levels when feedback and feedforward control is applied (March 13th 2005)

Finally, Model Predictive Control is applied in which the constraints as given in Paragraph 6.1.3 are taken into account. This Model Predictive Controller is actually implemented at the central operating room of the water board of Delfland. It advises the operators on the control actions to be taken over the next 24 hours. The objective function that is minimized is given in Formula 6.9:

$$\min_{\Delta Q_c, u^*} J =$$

$$\sum_{i=0}^{n} \left\{ e(k+i|k)^T \cdot Q_e \cdot e(k+i|k) \right\} +$$

$$\sum_{i=0}^{n} \left\{ \Delta e(k+i|k)^T \cdot Q_{\Delta e} \cdot \Delta e(k+i|k) \right\} +$$

$$\sum_{i=0}^{n} \left\{ e^*(k+i|k)^T \cdot Q_{e^*} \cdot e^*(k+i|k) \right\} + \qquad \text{Formula 6.9}$$

$$\sum_{i=0}^{n} \left\{ u^*(k+i|k)^T \cdot R_{u^*} \cdot u^*(k+i|k) \right\} +$$

$$\sum_{i=0}^{n-1} \left\{ \Delta Q_c(k+i\ k)^T \cdot R_{\Delta Q_c} \cdot \Delta Q_c(k+i|k) \right\}$$

where Q_e, $Q_{\Delta e}$ and Q_{e^*} represent the relative penalties on the states and $R_{\Delta Q_c}$ and R_{u^*} represent the relative penalties on the inputs. Table 6.2 gives the parameters used in the controller. The penalty values are derived from the method as described in Paragraph 3.2. The Maximum Allowed Value Estimate for e, Δe, e^*, ΔQ_c and u^* are 0.2 m, 0.05 m, 2.0e-3 m, 10 m^3/s and 1.0e6 m, respectively. These values are found without extensive tuning, but instead are based on estimates of their physical range.

Table 6.2 Parameters of Model Predictive Controlled water system

Parameter	Value	Unit
n	97	Steps (24 h)
Q_e	25	-
$Q_{\Delta e}$	400	-
Q_{e^*}	2.5e5	-
$R_{\Delta Q_c}$	1.0e-2	-
R_{u^*}	1.0e-12	-

Figure 6.13 and Figure 6.14 show the discharge of the pump stations and the resulting water levels when Model Predictive Control is applied.

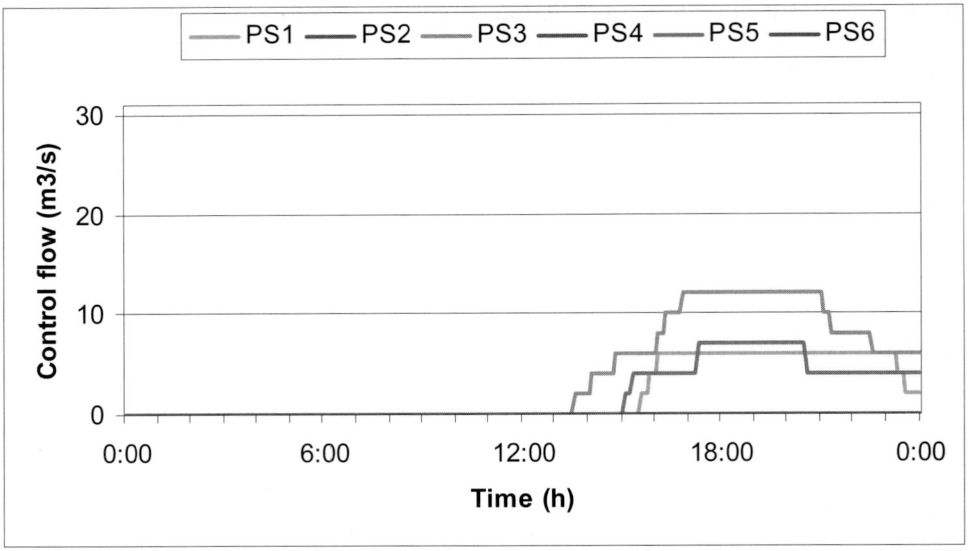

Figure 6.13 Discharge of pump stations when Model Predictive Control is applied (March 13th 2005)

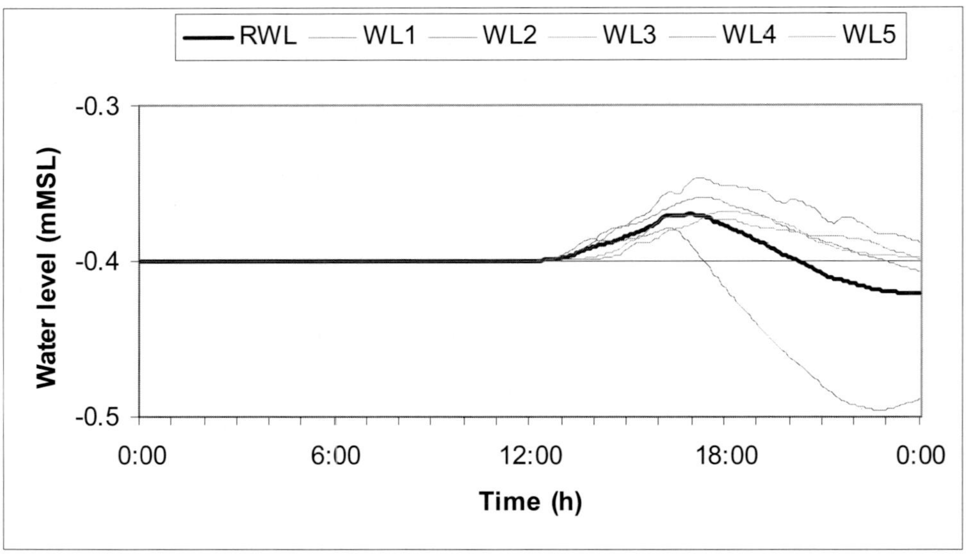

Figure 6.14 Representative water level and distributed water levels when Model Predictive Control is applied (March 13th 2005)

A different period is selected to show the functioning of MPC on extreme storm events. The storm event at September 13th and September 14th 1998 is notorious as it caused inundation on a large scale, resulting in approximately 500 million Euros of damage in the area of the water board of Delfland. The measured precipitation is given in Figure 6.15. Unfortunately, no forecast is available for this period. Therefore the actual precipitation is also used as forecast. Figure 6.16 shows the disturbance inflow into the storage canals.

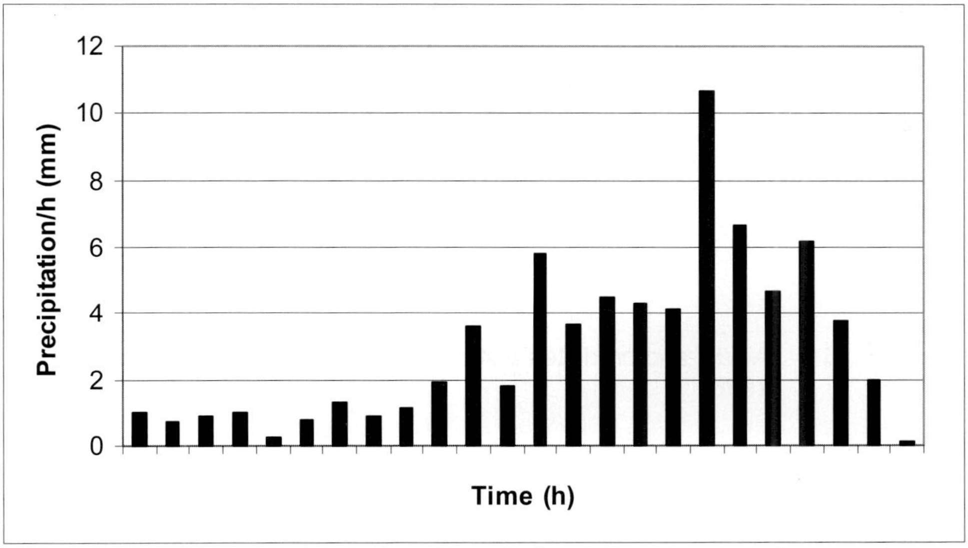

Figure 6.15 Precipitation of September 13th and September 14th 1998

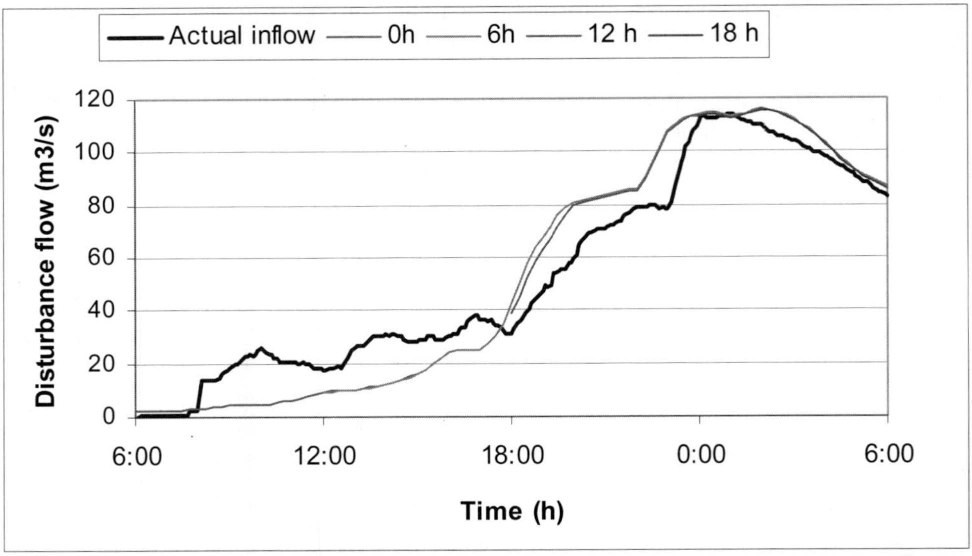

Figure 6.16 Actual and predicted disturbance (runoff) flow (September 13th 1998)

Figure 6.17 to 6.22 show the result of the heavy storm event at September 1998 when, respectively, feedback, feedback in combination feedforward and MPC are applied.

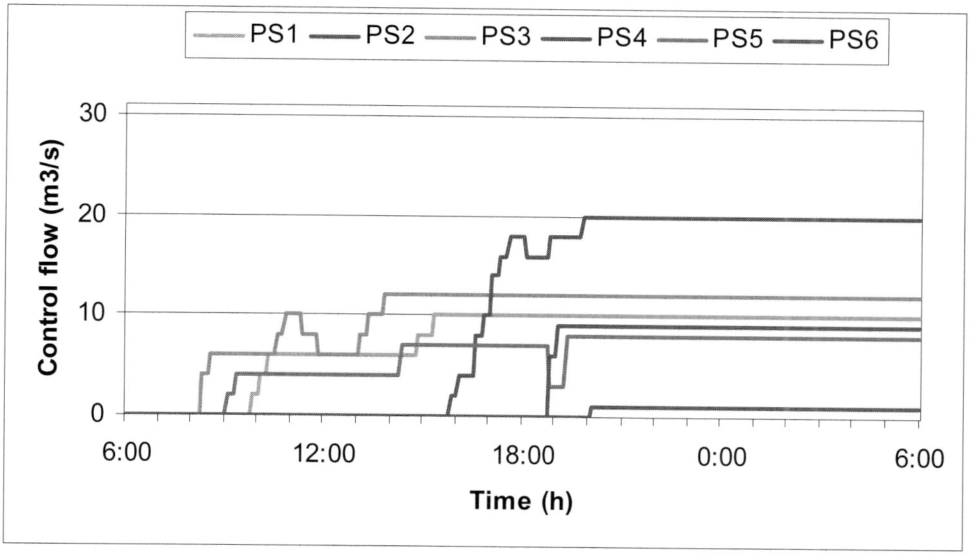

Figure 6.17 Discharge of pump stations when feedback control is applied (September 13th 1998)

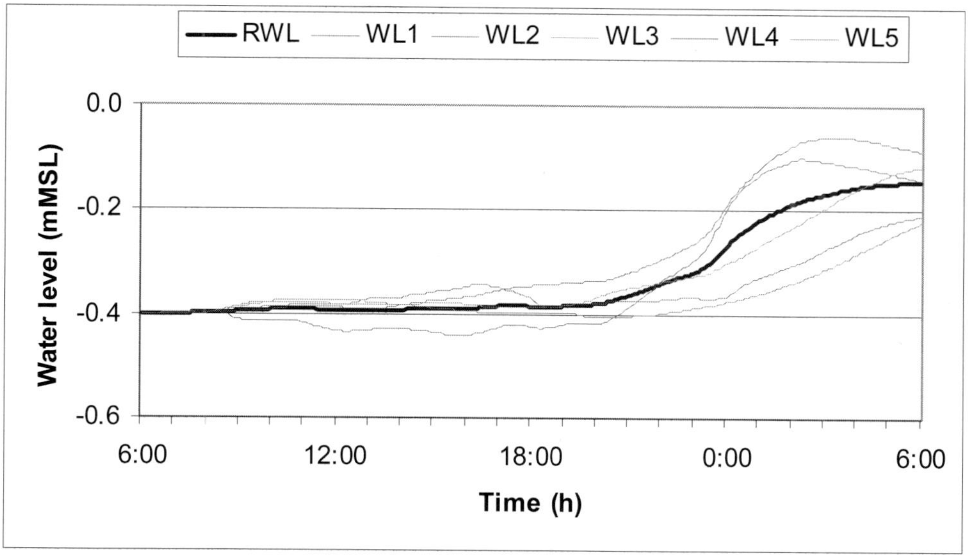

Figure 6.18 Representative water level and distributed water levels when feedback control is applied (September 13th 1998)

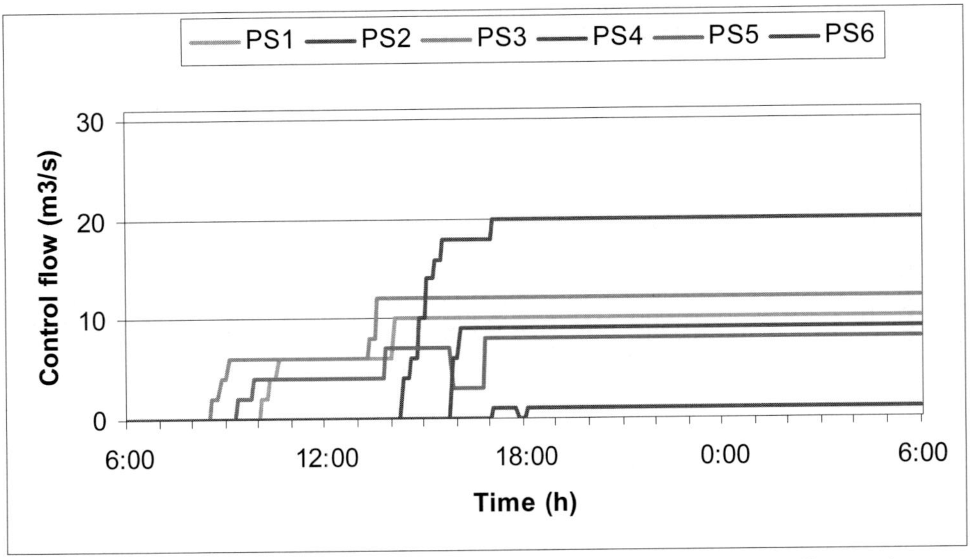

Figure 6.19 Discharge of pump stations when feedback and feedforward control is applied (September 13th 1998)

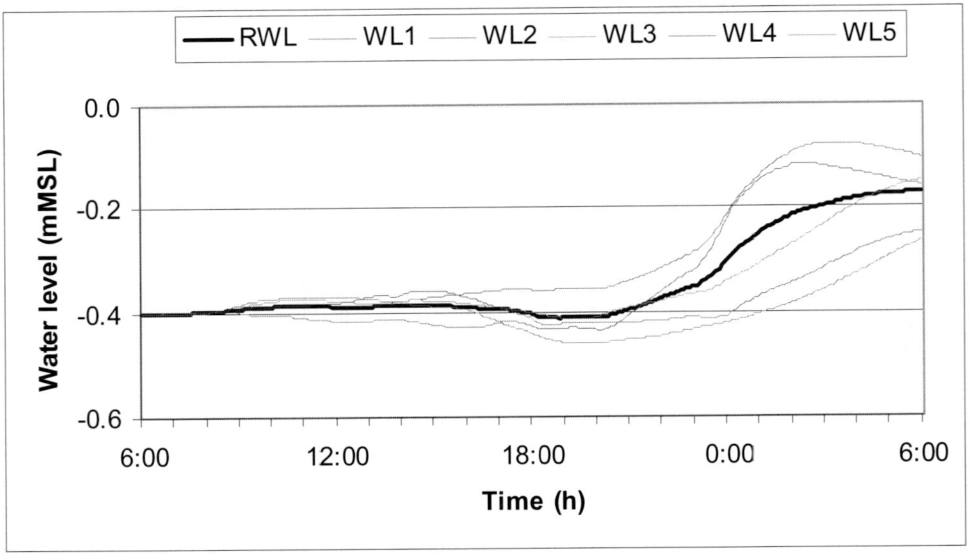

Figure 6.20 Representative water level and distributed water levels when feedback and feedforward control is applied (September 13th 1998)

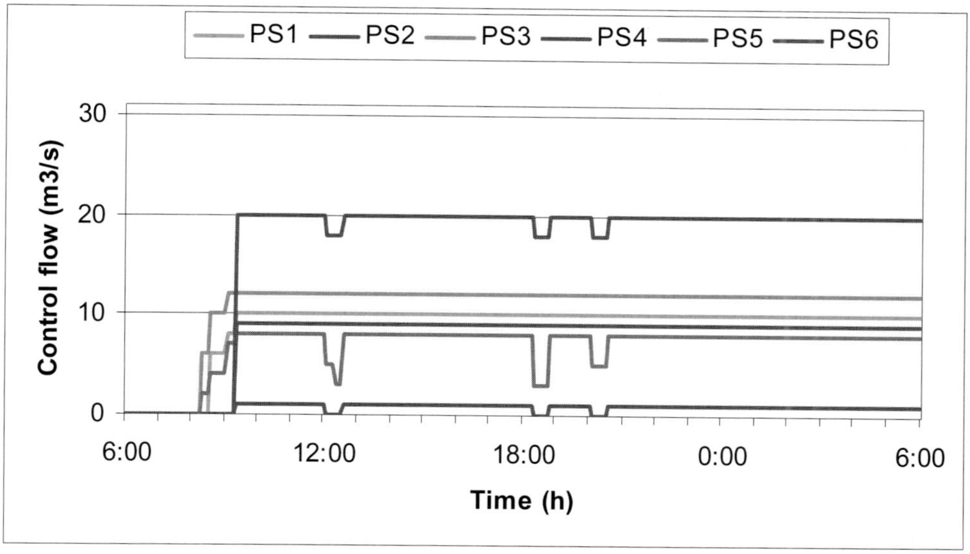

Figure 6.21 Discharge of pump stations when Model Predictive Control is applied (September 13th 1998)

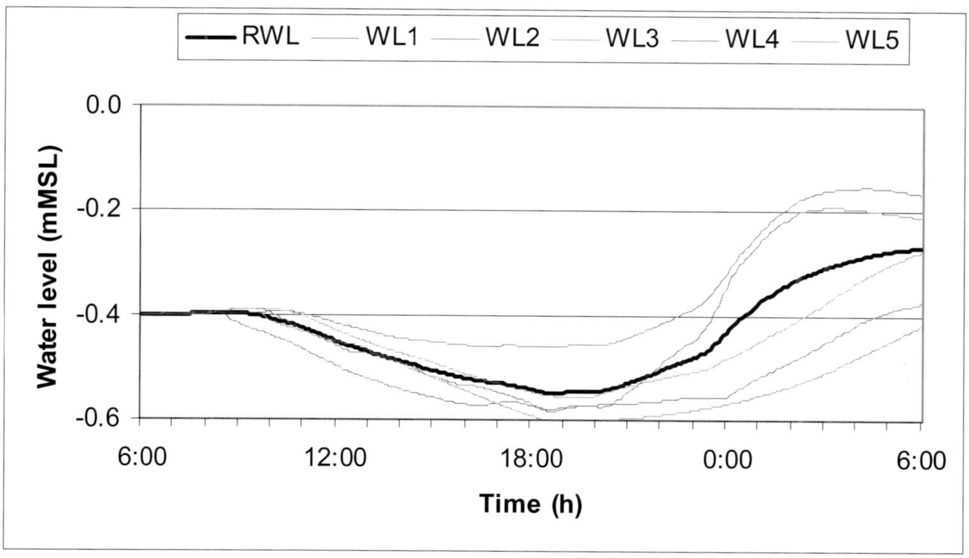

Figure 6.22 Representative water level and distributed water levels when Model Predictive Control is applied (September 13th 1998)

These simulations of the storm event at September 1998 are carried out based on the assumption that the actual precipitation is known in advance and can be used as forecast. This is not in concordance with reality. Especially heavy storm events are generally underestimated by the forecast. An evaluation of the storm event by the water board of Delfland shows that 30 mm of precipitation was expected while 72 mm fell. Next, the test with MPC is repeated with the prediction taken the same shape as the actual precipitation, but by using a multiplication factor of 30/72. Figure 6.23 shows the disturbance used and Figure 6.24 and Figure 6.25 show the simulation results.

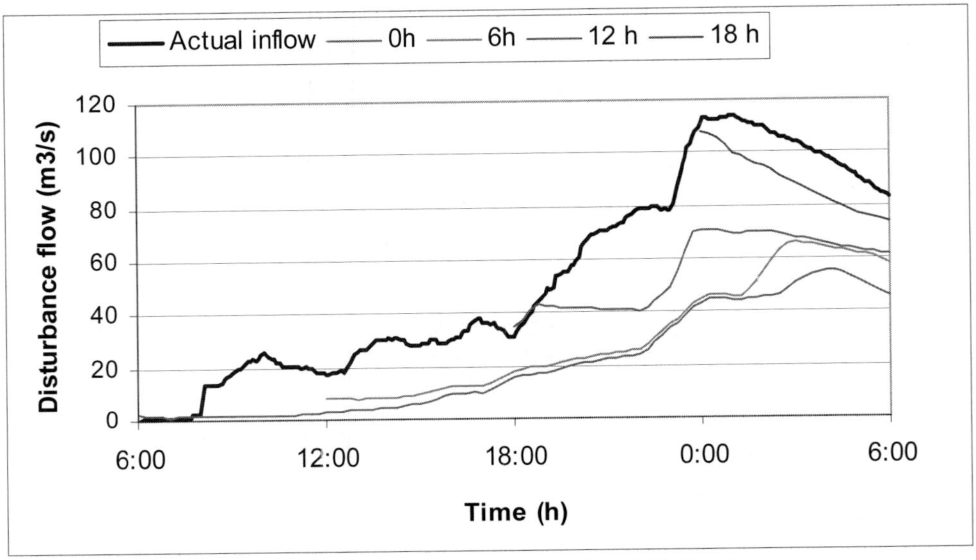

Figure 6.23 Actual and predicted disturbance (runoff) flow by using multiplication factor 30/72 (September 13th 1998)

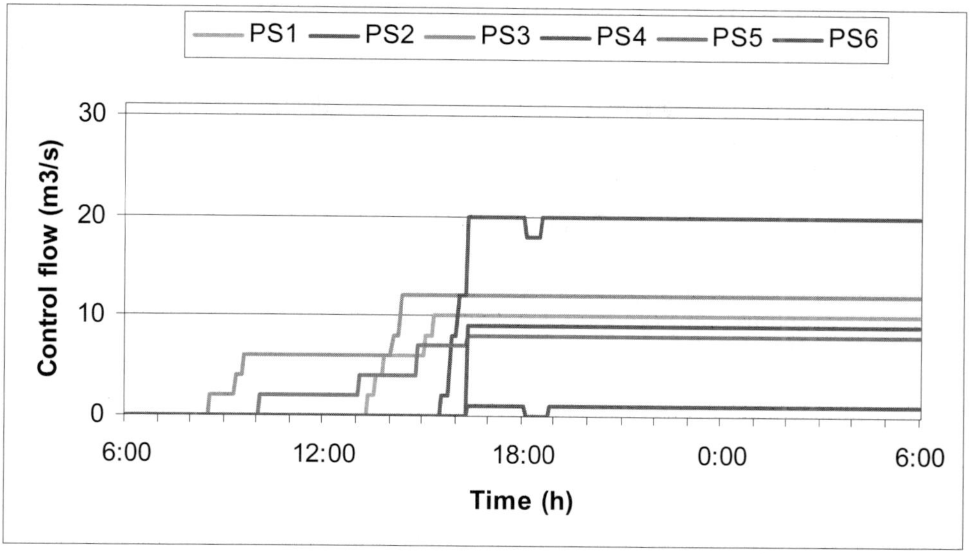

Figure 6.24 Discharge of pump stations when Model Predictive Control is applied by using multiplication factor 30/72 (September 13th 1998)

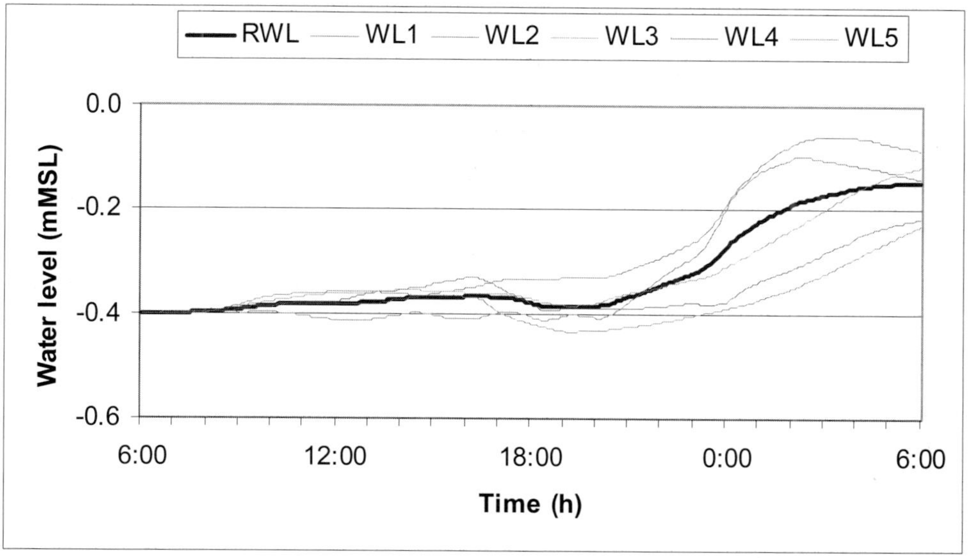

Figure 6.25 Representative water level and distributed water levels when Model Predictive Control is applied by using multiplication factor 30/72 (September 13th 1998)

6.1.5 Discussion on results of control on drainage system Delfland

The test of March 13th 2005 is not a critical situation. All controllers keep the water level within the allowed range. No constraints are violated, so applying Model Predictive Control does not provide extra functionalities compared to feedback and feedforward control. For the heavy storm event at September 13th and 14th the situation is severe and for the controlled water system a high performance is required. The behavior of the different controllers is explained below.
The feedback controller only reacts when there is actually a deviation from setpoint. Therefore, this controller is always delayed in its reaction on the disturbance inflow. This is not a problem when the disturbance flow remains lower than the total pump capacity. If the feedback controller is tuned well, the representative water level will generally remain within the allowed range (see Figure 6.8).
An improvement to the feedback controller can be made by adding the inverse of the effect that the disturbance inflow has on the water level as feedforward signal. The addition of the feedforward component makes the pumps start earlier (see Figure 6.7 and Figure 6.11). In case the disturbance inflow remains lower than the total pump capacity and the disturbance can be accurately estimated, the representative water level will remain close to setpoint (see Figure 6.12).
When the disturbance inflow remains lower than the total pump capacity, the Model Predictive Controller functions comparable to the feedback controller in combination with feedforward control (see Figure 6.13 and Figure 6.14). As the penalty on the change in control flow is not high, MPC can keep the objective function low by keeping the deviation from setpoint small. In case the disturbance inflow exceeds the total pump capacity, the water levels in the storage canals will start to rise (see Figure 6.18 and Figure 6.20). To keep the value of the objective function low, MPC will lower the representative water level before the disturbance inflow becomes too high. This can be observed from the early switching on of the pump stations even before the storm event actually takes place (see Figure 6.21). This results in a 15 centimeters lower peak in the water level. Nevertheless, at present the forecast of the precipitation of extreme storm events is not accurate enough to achieve such an improvement by applying Model Predictive Control. For the moment, improvements of a few centimeters are more realistic (see Figure 6.25). For the water board of Delfland this situation will improve considerably when radar images of the lower atmosphere are incorporated into the short term forecast. These images can accurately predict the local precipitation over the next 3 to 6 hours. For the fast system of Delfland this time frame is long enough to anticipate the disturbance by lowering the water levels.

6.2 *Local control of irrigation canal reach Delta Mendota*

The Delta-Mendota Canal is located on the West side of the San Joaquin Valley in California. It measures about 200 kilometers. The canal is essential for irrigation supply as part of the San Luis Unit and the Central Valley Project Delta Division. The canal starts southwest of Stockton (about 100 km east of San Francisco), where it receives water pumped at the Tracy Pumping Plant from the Sacramento-San Joaquin River Delta. It flows through the western part of the San Joaquin Valley to the Mendota Pool, approximately 50 kilometers west of Fresno. Bruggers (2004) provides a full description of the canal. Figure 6.26 shows the location of the Delta Mendota Canal.

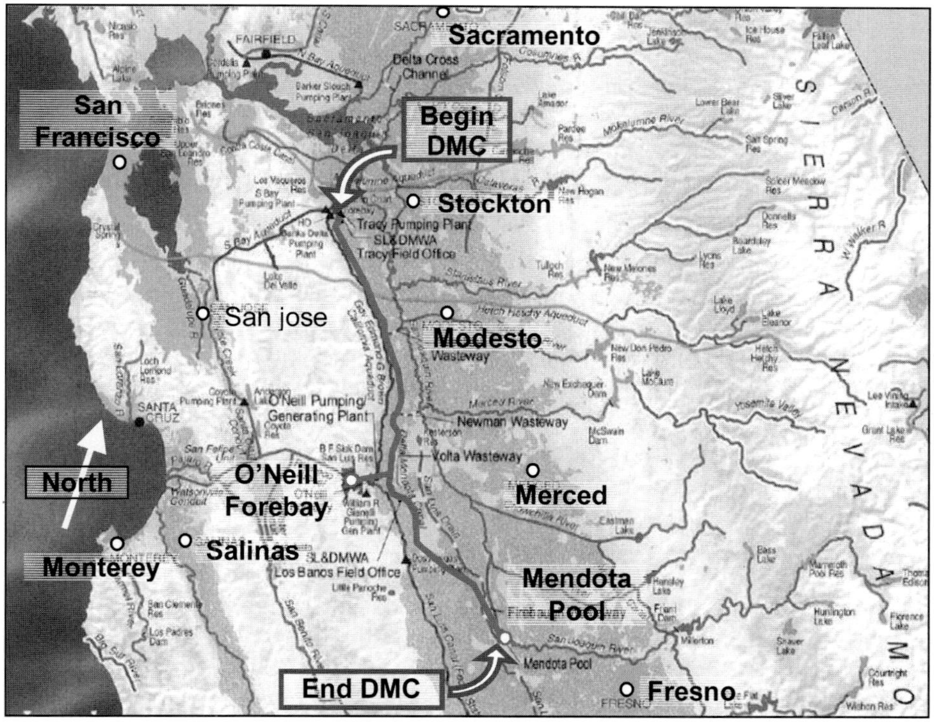

Figure 6.26 Map of Delta Mendota Canal

The canal consists of 8 reaches in series divided by check structures. These check structures are submerged undershot gates. The water levels upstream of each gate are locally controlled at a fixed set point. Figure 6.27 shows a longitudinal profile of the canal.

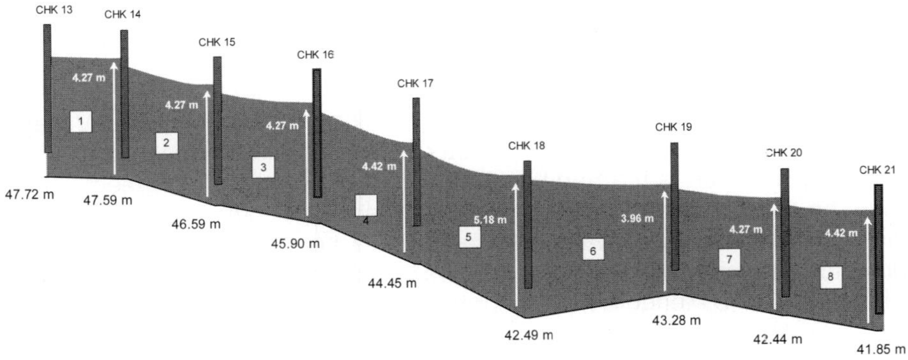

Figure 6.27 Longitudinal profile of Delta Mendota Canal

6.2.1 Objective of local control of irrigation canal reach Delta Mendota

The objective of the controller on the Delta Mendota Canal is to transport water from the upstream side to the turnouts that are located all along the canal. In case extra water is needed, the head gate is adjusted and the water flows towards the first gate in the canal. Here, the local controller will open the gate to correct for the deviation from setpoint caused by the extra water. The water flows to the next gate that will open and so on. This is referred to as local upstream control.
The reason for setting up the control of the canal as local control is that past experiences have proven communication lines between sites to be a weak link. Therefore, the communication lines are kept out of the feedback control loop. However, the communication lines are available along the canal and can be used occasionally. Given this restriction, the objective of the controlled water system is to keep the water levels as close to setpoint as possible. Since the gates are heavy, the small deviations need to be achieved with as few adjustments to the gates as possible.
The present controllers cause the water levels to have a certain deviation given the step change in flow. If the performance of these controllers can be improved, larger flow change steps can be used with comparable water level deviations and the delivery of water through the canal becomes faster.

6.2.2 Characteristics of local control of irrigation canal reach Delta Mendota

The controllers that are applied are Proportional Integral Filter controllers (Schuurmans, J. (1997)). Use is made of a first order low-pass filter to attenuate the amplification of the critical resonance wave in the canal:

$$\Delta Q_c(k) = K_p \cdot \Delta e_f(k) + K_i \cdot e_f(k)$$ Formula 6.10

$$e_f(k) = F_c \cdot e_f(k-1) + (1 - F_c) \cdot e(k)$$ Formula 6.11

$$e(k) = h(k) - h_{ref}$$ Formula 6.12

$$\Delta Q_c(k) = Q_c(k) - Q_c(k-1)$$ Formula 6.13

$$\Delta e(k) = e(k) - e(k-1)$$ Formula 6.14

where Q_c represents the control flow (m^3/s), K_p the proportional gain, K_i the integral gain, F_c the filter factor, e_f the filtered error between water level and setpoint (m), e the measured error (m), h the water level upstream of the gate (m), h_{ref} the setpoint for the water level (m) and k the discrete time step index,. The gate openings are calculated with a flow controller (Schuurmans, J. (1997)) by inverting the submerged undershot gate flow given in Formula 2.10:

$$h_g(k) = h_{cr} + \frac{Q_c(k)}{C_g \cdot W_g \cdot \mu_g \cdot \sqrt{2 \cdot g \cdot (h_1(k) - h_2(k))}}$$ Formula 6.15

where h_g represents the gate height (mMSL), h_{cr} the crest height (mMSL), C_g the calibration coefficient (=1.0), W_g the gate width (=18.3 m), μ_g the contraction coefficient (=0.63), g the gravitational acceleration (=9.81 m/s^2), h_1 the upstream water level (mMSL) and h_2 the downstream water level (mMSL). Both the controller and the flow controller have a control time step T_c of 60 seconds.
The controller parameters are tuned with an optimization program that gives a high performance and avoids instability at both low and high flow (Overloop et al. (2005b)). This program uses the linear models of the low and high flow steady solution characterized by the delay time, storage area of the back water, sensitivity to the critical resonance wave and the frequency of this resonance wave. The optimized control parameters should function well at both working points. This means that the performance of the controlled water system is by definition lower than if the optimization was carried out at only one working point. If a feedforward controller that uses the communication lines occasionally is added to the local feedback control, the performance will improve.
An additional improvement can be achieved by replacing the filter for a controller that actively reacts on the (resonance) waves in the canal. This requires an internal model that can describe the waves in a canal. It is known that the De Saint Venant equations can describe the waves accurately (Chow (1959)). A model that uses these non-linear partial differential equations can simulate the

water flows and water levels over the entire range of working points (Stelling & Booij (1994)). In Arnold et al. (1998) and (1999) an implementation of a model based controller on a canal in Germany is presented that uses a De Saint Venant internal model.
To test the different controllers only the first reach is examined. This allows for a fast, transparent and isolated test environment without any interactions with other subsystems. Also the test is setup to analyze the difference in performance between the control types and not to design a completely new control system for the canal. The outcome of the tests can be extrapolated to the situation for all reaches. The controlled reach is shown in Figure 6.28.

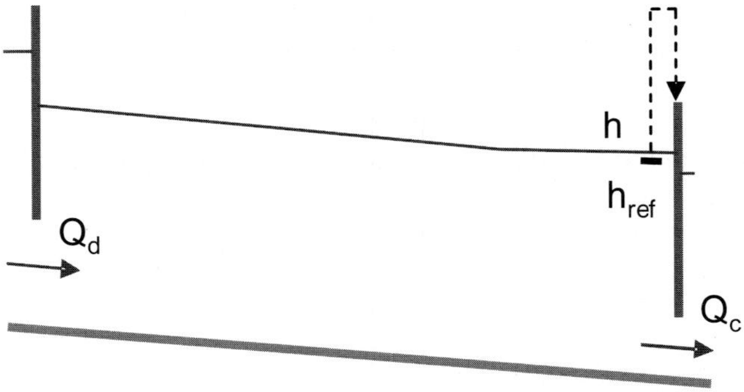

Figure 6.28 Longitudinal profile of first canal reach of Delta Mendota Canal

The parameters that are used in the control of reach 1 are given in Table 6.3

Table 6.3 Parameters of controlled reach 1

Parameter	Value	Unit
T_c	60	s
h_{ref}	51.86	mMSL
K_p	434	-
K_i	22	-
F_c	0.74	-

6.2.3 Constraints of local control of irrigation canal reach Delta Mendota

The constraints applied to the control of the first reach of the Delta Mendota Canal are:

- The control flow must be higher than 0 m^3/s;
- The control flow must be lower than 80 m^3/s.

6.2.4 Results of local control of irrigation canal reach Delta Mendota

The following controllers are tested on a test of 10 m^3/s step change at 20 m^3/s, 40 m^3/s and 60 m^3/s base flow and a shut down test:

- Present local Proportional Integral Filter controller;
- Present local Proportional Integral Filter controller with feedforward signal of upstream flow change;
- Sequential configuration of Model Predictive Controller that uses the local water level and the feedforward signal of the upstream flow change.

The controller, including the feedforward controller, is an extension of the feedback controller with a prediction of the disturbance ΔQ_d. The feedforward controller functions according to:

$$\Delta Q_c(k) = K_p \cdot \Delta e_f(k) + K_i \cdot e_f(k) + \Delta Q_d(k - k_d) \qquad \text{Formula 6.16}$$

$$\Delta Q_d(k - k_d) = Q_d(k - k_d) - Q_d(k - k_d - 1) \qquad \text{Formula 6.17}$$

where Q_d represents the disturbance flow at the upstream side of the reach (m^3/s) and k_d the number of delay time steps from upstream side to downstream side of the reach. The number of delay steps is the rounded delay time at average flow divided by the control time step. This results in a value for k_d equal to 17 delay steps.

The sequential configured Model Predictive Controller uses an internal model of five water level solutions along the first reach with a spatial discretization Δx of 1420 m. At these locations, the De Saint Venant equations are linearized according to Stelling & Duinmeyer (2003). The flow between two water level locations l and l+1 is:

$$Q_l(k) = A_{f,l}(k) \cdot \left(fu_l(k) \cdot \left(h_l(k) - h_{l+1}(k) \right) + ru_l(k) \right) \qquad \text{Formula 6.18}$$

where *fu* and *ru* represent the time-variant parameters calculated from the momentum equations including inertia, advection, water level gradient and bed friction (see Formula 2.2). A_f represents the time-variant wetted area calculated from the upstream water level, bottom width and the side slope. The state space model that gives the continuity equations at time step k is presented in Formula 6.19 (index k of the parameters is left out for readability). This formula is presented to show the band matrix structure of the matrices:

$$
\begin{bmatrix} h_1(k+1) \\ h_2(k+1) \\ h_3(k+1) \\ h_4(k+1) \\ h_5(k+1) \\ Q(k) \end{bmatrix} =
$$

$$
\begin{bmatrix}
1-\frac{T_c \cdot A_{f,1} \cdot fu_1}{A_{s,1}} & \frac{T_c \cdot A_{f,1} \cdot fu_1}{A_{s,1}} & 0 & 0 & 0 & 0 \\
\frac{T_c \cdot A_{f,1} \cdot fu_1}{A_{s,2}} & 1-\frac{T_c \cdot A_{f,1} \cdot fu_1}{A_{s,2}}-\frac{T_c \cdot A_{f,2} \cdot fu_2}{A_{s,2}} & \frac{T_c \cdot A_{f,2} \cdot fu_2}{A_{s,2}} & 0 & 0 & 0 \\
0 & \frac{T_c \cdot A_{f,2} \cdot fu_2}{A_{s,3}} & 1-\frac{T_c \cdot A_{f,2} \cdot fu_2}{A_{s,3}}-\frac{T_c \cdot A_{f,3} \cdot fu_3}{A_{s,3}} & \frac{T_c \cdot A_{f,3} \cdot fu_3}{A_{s,3}} & 0 & 0 \\
0 & 0 & \frac{T_c \cdot A_{f,3} \cdot fu_3}{A_{s,4}} & 1-\frac{T_c \cdot A_{f,3} \cdot fu_3}{A_{s,4}}-\frac{T_c \cdot A_{f,4} \cdot fu_4}{A_{s,4}} & \frac{T_c \cdot A_{f,4} \cdot fu_4}{A_{s,4}} & 0 \\
0 & 0 & 0 & \frac{T_c \cdot A_{f,4} \cdot fu_4}{A_{s,5}} & 1-\frac{T_c \cdot A_{f,4} \cdot fu_4}{A_{s,5}} & -\frac{T_c}{A_{s,5}} \\
0 & 0 & 0 & 0 & 0 & 1
\end{bmatrix} \cdot
$$

$$
\begin{bmatrix} h_1(k) \\ h_2(k) \\ h_3(k) \\ h_4(k) \\ h_5(k) \\ Q_c(k-1) \end{bmatrix} +
$$

$$
\begin{bmatrix} 0 \\ 0 \\ 0 \\ 0 \\ 0 \\ 1 \end{bmatrix} \cdot [\Delta Q_c(k)] +
\begin{bmatrix}
\frac{T_c}{A_{s,1}} & 0 & 0 & 0 & 0 & \frac{T_c}{A_{s,1}} \\
0 & \frac{T_c}{A_{s,2}} & 0 & 0 & 0 & 0 \\
0 & 0 & \frac{T_c}{A_{s,3}} & 0 & 0 & 0 \\
0 & 0 & 0 & \frac{T_c}{A_{s,4}} & 0 & 0 \\
0 & 0 & 0 & 0 & \frac{T_c}{A_{s,5}} & 0 \\
0 & 0 & 0 & 0 & 0 & 0
\end{bmatrix} \cdot
\begin{bmatrix}
-A_{f,1} \cdot ru_1 \\
A_{f,1} \cdot ru_1 - A_{f,2} \cdot ru_2 \\
A_{f,2} \cdot ru_2 - A_{f,3} \cdot ru_3 \\
A_{f,3} \cdot ru_3 - A_{f,4} \cdot ru_4 \\
A_{f,4} \cdot ru_4 \\
Q_d(k)
\end{bmatrix}
$$

Formula 6.19

All parameters in this model are time-variant and are calculated in an iterative loop from the water level and flows at time step index *k* and *k*+1 with a weight factor of ξ=0.5 comparable to the Crank Nicolson method (Stelling & Booij (1994)). For that reason, Model Predictive Control in sequential configuration is used rather than standard Model Predictive Control.

The objective function that is used is:

$$
\min_{\Delta Q_c} J = \sum_{i=0}^{n} \left\{ e_5(k+i|k)^T \cdot Q_e \cdot e_5(k+i|k) \right\} + \sum_{i=0}^{n-1} \left\{ \Delta Q_c(k+i|k)^T \cdot R_{\Delta Q_c} \cdot \Delta Q_c(k+i|k) \right\}
$$

Formula 6.20

where Q_e represents the relative penalty on the water level deviation at the downstream side of the reach (upstream of the control gate) and $R_{\Delta Qc}$ the relative penalty on the change in control flow of the downstream gate. Table 6.4 gives the parameters used in the controller. The penalty values are derived from the method as described in Paragraph 3.2. The Maximum Allowed Value Estimate for

e and ΔQ_c are 0.2 m and 10 m^3/s, respectively. These values are found without extensive tuning, but are based on estimates of the physical range of regular operating points.

Table 6.4 Parameters of Model Predictive Control DMC

Parameter	Value	Unit
n	61	Steps (1 h)
Q_e	25	-
$R_{\Delta Q_c}$	0.01	-

The model is updated from the measurement of the water level upstream of the control gate with a standard Kalman filter (Kalman (1960)).The tuning of the Kalman filter takes place at the beginning of every control time step from the state space model of the previous solution and the standard deviations of the uncertainty on the modeled water levels, modeled control flow and measurement of the water level taken as 0.02 m, 0.1 m^3/s and 0.1 m, respectively. The resulting Kalman feedback $\boldsymbol{L}$ matrix as described in Paragraph 3.1 of the first simulation step is:

$$\boldsymbol{L} = \begin{bmatrix} 0.0286 \\ 0.0334 \\ 0.0454 \\ 0.0758 \\ 0.1632 \\ -0.9127 \end{bmatrix}$$

The first five values correct the five water level deviation states, while the last corrects the discharge state of the inflow at the previous time step. The first four values in the vector are low and slightly increasing, while the fifth value is higher. This can be explained by the location of the measurement. This measurement location coincides with this last modeled water level deviation. The last value in the vector is relatively low as it corrects the discharge, while the others correct on the water level deviation. The water level deviation e is only a few centimeters, while the discharge Q_c is 40 m^3/s.

To test how well the simplified internal model resembles a more accurate description of the canal, a step test is carried out with both the simplified model and an accurate model. The simplified model (Δx=1420 m, ΔT=60 s) is programmed in Matlab (MathWorks (1992)) and the accurate model (Δx=100 m, ΔT=10 s) is build in an accurate hydro-dynamic model (Sobek (2000)). The base flow is 40 m^3/s. At the same time, a step-wise flow change of 10 m^3/s is made to the upstream gate and to the downstream gate. Figure 6.29 shows the water level upstream in the reach of the accurate model, together with the predicted water level over the prediction horizon of the simplified model at 30, 60 and 90 minutes, respectively. Figure 6.30 is the same representation, but now of the (measured) water level downstream in the reach.

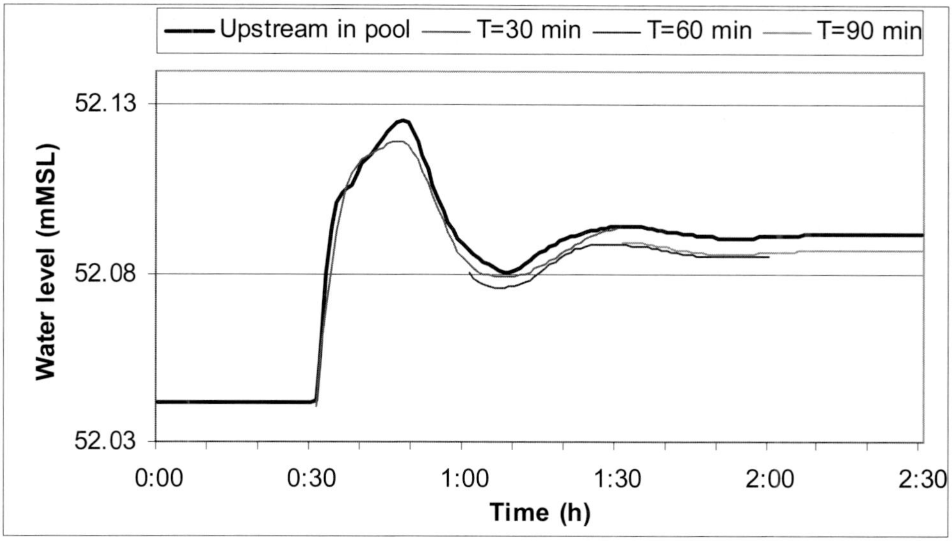

Figure 6.29 Water levels at upstream side of the reach resulting from step in flows of gates for accurate and simplified model

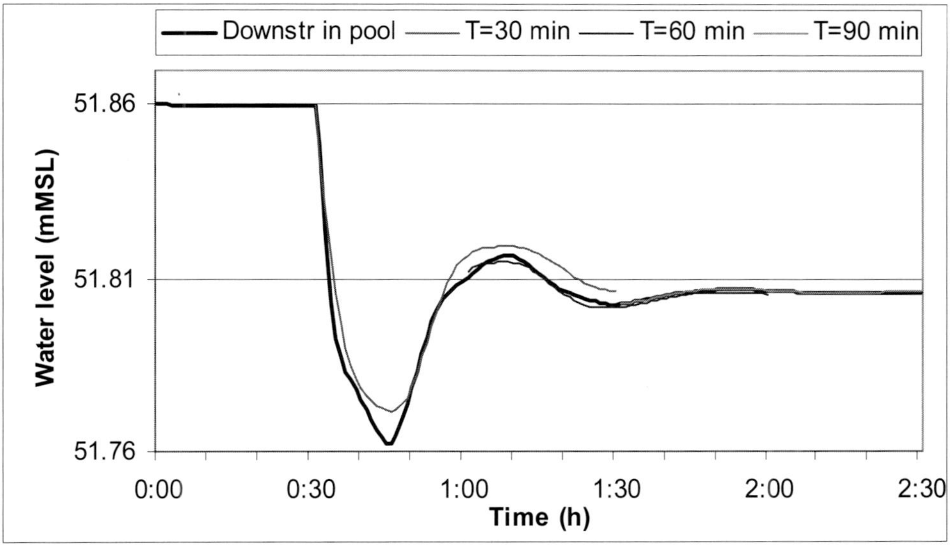

Figure 6.30 Water levels at downstream side of the reach resulting from step in flows of gates for accurate and simplified model

These results show that the simplified model is able to accurately predict the basic dynamics of the water levels and flows in the reach. Clearly, the basic frequency resonance wave is captured very well. The dynamics with higher

frequencies are predicted less accurately. This can be explained by the limited number of spatial discretization points of the simplified model. By adding more points, the simplified model will describe the higher harmonics better. An additional conclusion is that the update procedure functions very well for the measured location. This can be seen in Figure 6.30 where the start of every prediction is very close to the actual water level. As the predicted water levels as presented in Figure 6.29 are at the other side of the reach, the update procedure functions less, but still reasonable.
In Figure 6.31 to 6.36 the results of the three controllers are presented on a step-wise change in flow at the upstream gate of 10 m^3/s at base flows of 20 m^3/s, 40 m^3/s and 60 m^3/s, respectively.

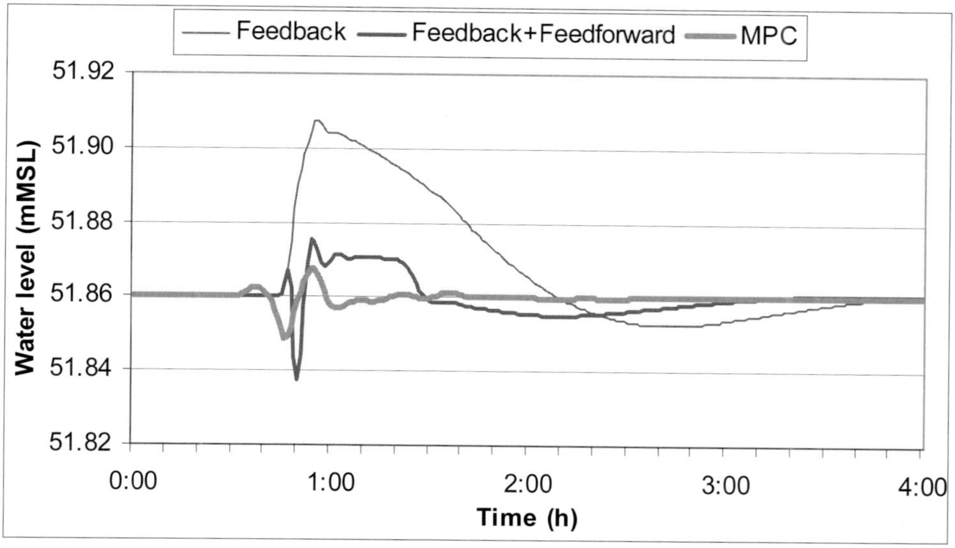

Figure 6.31 Controlled water level of three controllers (Step = 10 m^3/s, Base flow = 20 m^3/s)

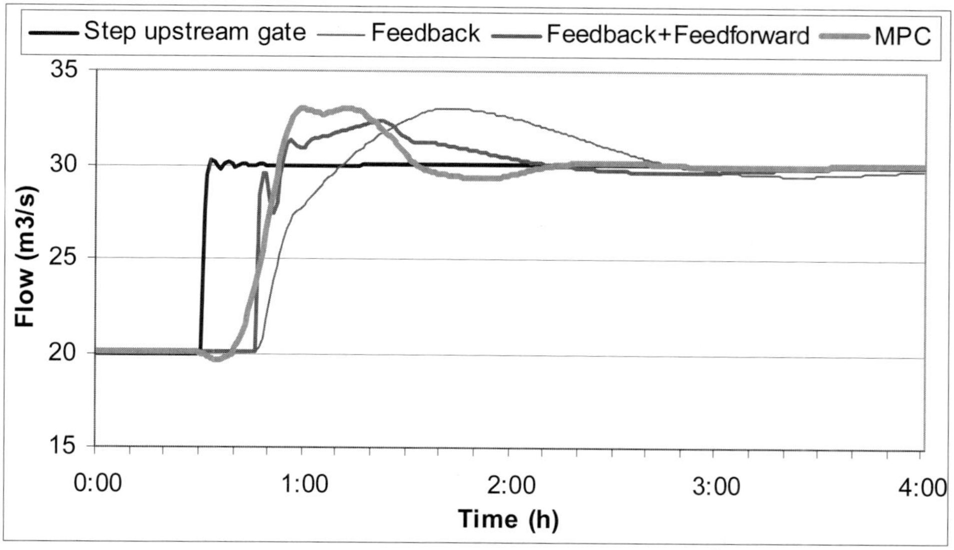

Figure 6.32 Control flow of three controllers (Step = 10 m^3/s, Base flow = 20 m^3/s)

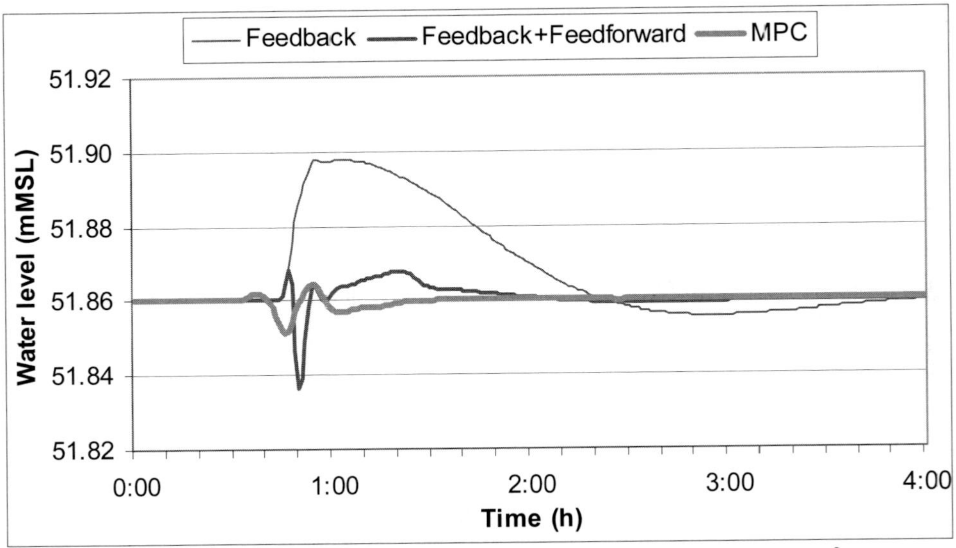

Figure 6.33 Controlled water level of three controllers (Step = 10 m^3/s, Base flow = 40 m^3/s)

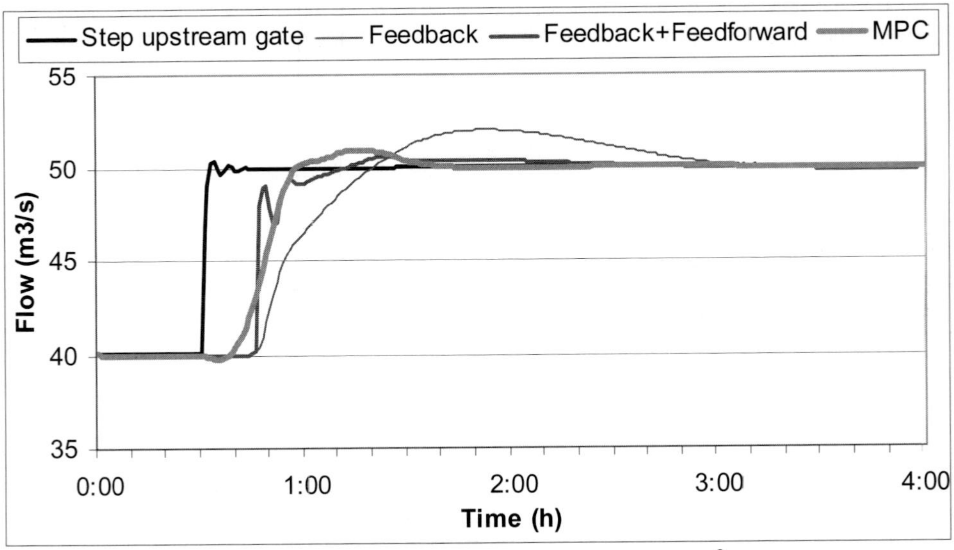

Figure 6.34 Control flow of three controllers (Step = 10 m^3/s, Base flow = 40 m^3/s)

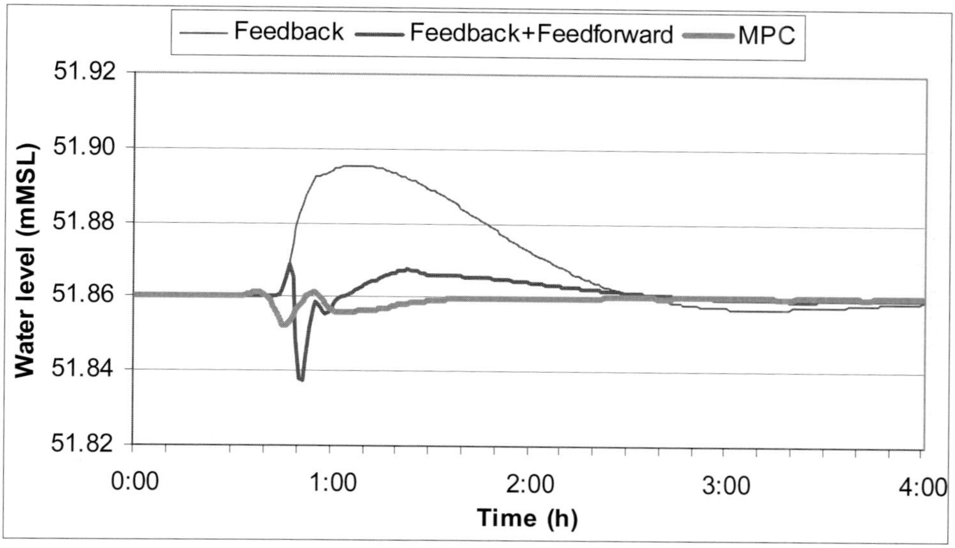

Figure 6.35 Controlled water level of three controllers (Step = 10 m^3/s, Base flow = 60 m^3/s)

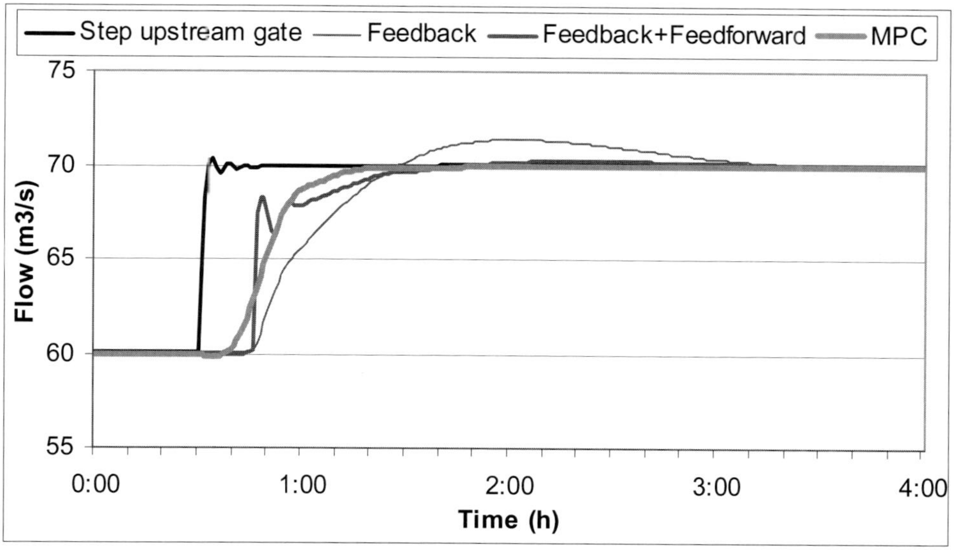

Figure 6.36 Control flow of three controllers (Step = 10 m^3/s, Base flow = 60 m^3/s)

Finally, an extreme event is tested in which the upstream gate closes completely, for example when the lifting installation of the gate breaks down. Again, this shut down test is performed by the three controllers and the results are presented in Figure 6.37 and 6.38.

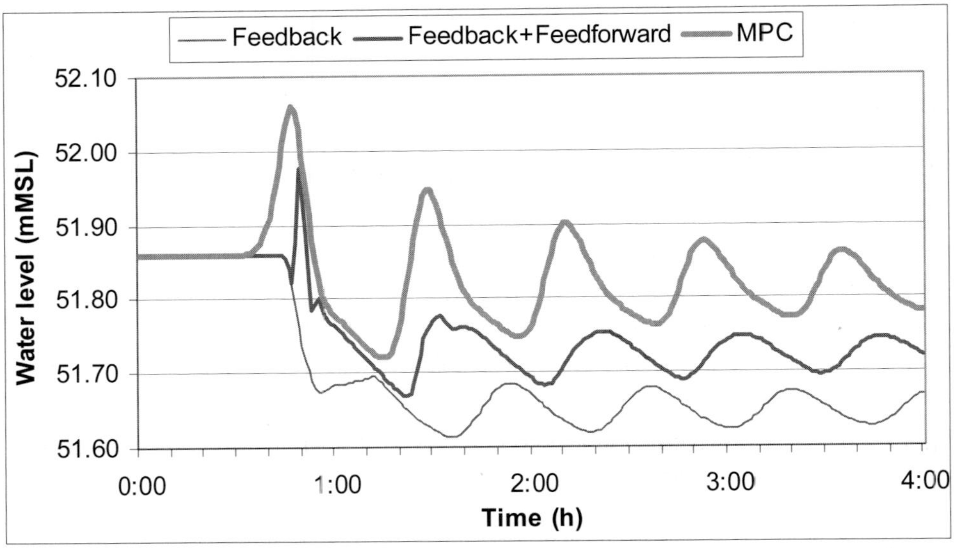

Figure 6.37 Controlled water level of three controllers during shut down by failure of upstream gate

The almost undamped oscillation is the resonance wave corresponding with the basic frequency of the canal reach. This canal reach is flat, short and deep, so according to the theory in Figure 2.7, sensitive to these reflecting waves.

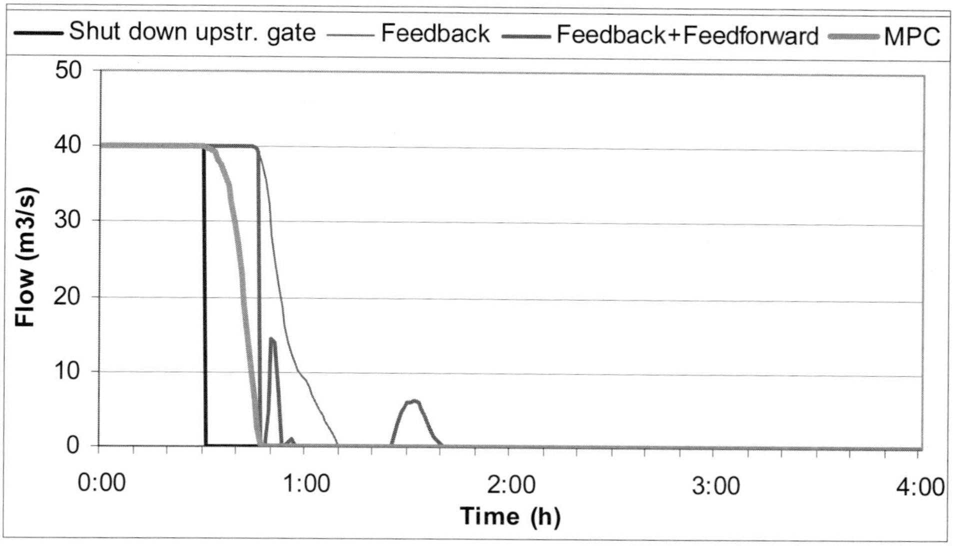

Figure 6.38 Control flow of three controllers during shut down by failure of upstream gate

6.2.5 Discussion on results of local control of irrigation canal reach Delta Mendota

It is clear from Figures 6.31 to 6.36 that the sequential configuration of Model Predictive Controller outperforms the other two controllers. The water level deviations are at least twice as small and so are the settling times. The control flow is much smoother and, consequently, so is the change in gate height. This shows that sequential configuration of MPC effectively controls disturbance waves in the reach.

The shut down test as presented in Figure 6.37 and 6.38 shows that sequential configured MPC can handle extreme situations and smoothly controls the water system along the most optimal trajectory (given the objective function and the constraints) to a new steady solution for the control gate. In the test, the water levels still oscillate with the basic frequency resonance wave until the end of the test. The amplitude of the oscillations of sequential configured MPC is even higher than for the other controllers. This is because the sequential configuration of MPC is able to maintain more volume in the reach, as it controls the water level closer to setpoint. As resonance waves are attenuated less in deeper reaches, the amplitude of the oscillations remains higher. Also, after the shut down of the control gate, sequential configured MPC is out of control and has no influence anymore on the oscillations. Note that, when the upstream gate is repaired and the system has to start up again, the higher water level in the sequential configured MPC test will enable a fast recovery of the controlled water system. As an estimate of all water levels along the reach is available in the sequential configuration of Model Predictive Control, the water level at a different location can be controlled, even though only the local water level upstream of the control gate is measured. This can be useful if the offtake structure is located at a different location in the reach. The configuration is still local control, but can control the water level at a distant location. Stretching this functionality even further, allows for volume control, by controlling both the water level at the upstream and at the downstream side in the reach. This can simply be done by using the objective function:

$$\min_{\Delta Q_c} J =$$

$$\sum_{i=0}^{n}\left\{e_1(k+i\,|\,k)^T\cdot Q_e\cdot e_1(k+i\,|\,k)+e_5(k+i\,|\,k)^T\cdot Q_e\cdot e_i(k+i\,|\,k)\right\}+$$

$$\sum_{i=0}^{n-1}\left\{\Delta Q_c(k+i\,|\,k)^T\cdot R_{\Delta Q_c}\cdot \Delta Q_c(k+i\,|\,k)\right\}$$

Formula 6.21

An even more accurate volume controller can be executed by using the objective function:

$$\min_{\Delta Q_c} J =$$

$$\sum_{i=0}^{n}\left\{\begin{matrix} e_1(k+i\,|\,k)^T\cdot Q_e\cdot e_1(k+i\,|\,k)+ \\ e_2(k+i\,|\,k)^T\cdot Q_e\cdot e_2(k+i\,|\,k)+ \\ e_3(k+i\,|\,k)^T\cdot Q_e\cdot e_3(k+i\,|\,k)+ \\ e_4(k+i\,|\,k)^T\cdot Q_e\cdot e_4(k+i\,|\,k)+ \\ e_5(k+i\,|\,k)^T\cdot Q_e\cdot e_i(k+i\,|\,k) \end{matrix}\right\}+$$

Formula 6.22

$$\sum_{i=0}^{n-1}\left\{\Delta Q_c(k+i\,|\,k)^T\cdot R_{\Delta Q_c}\cdot \Delta Q_c(k+i\,|\,k)\right\}$$

where e_1, e_2, e_3, e_4 and e_5 represent the predicted water level deviations from their setpoints at all discretization points of the internal model.

6.3 Off/on control of pump station IJmuiden

The 'Noordzeekanaal-Amsterdam Rijnkanaal' (NZK-ARK) is a large canal in the Western part of The Netherlands. It connects the lake Markermeer, the River Lek, the City of Amsterdam and the City of Utrecht to the North Sea. It is used for shipping, draining water from the adjacent water boards and cities and for keeping the salinity of the lake 'Markermeer' low enough. Furthermore, a large number of people are living in house boats on this canal. Figure 6.39 shows the NZK-ARK on a map of the Western part of The Netherlands (Weissenbruch (2003)).

Before the industrial era, drainage water was discharged only during low tide by gravity flow through the sluice gate at the end of the canal. During this period, the water level in the canal fluctuated more than one meter. Since electric motors in large pump installations became available, excessive water could be drained out of the canal during high tide. The installation of four parallel pumps with a total capacity of 160 m^3/s resulted in a more constant water level, allowing for a more intensive usage of the canal for various purposes.

The present management of the canal is executed in a central operating room at the end of the canal, where the sluice gate and the pump station are located. Here, the operators have measurements at their disposal of the water level in the canal, predictions of the sea water level over the next 24 hours, forecasts of the precipitation over the next days and inflow from the pump stations of the water boards. Based on their experience, they select the number of pumps and determine the operating hours these pumps need to run in addition to the gravity flow through the sluice.

In 2004, two extra pumps were installed to ensure the safety of this important part of The Netherlands, especially regarding the expected future rise of sea water level. The total pump capacity now measures 260 m^3/s. At the moment of finalizing this research, it is the largest pump station in Europe. Figure 6.40 shows one of the new pumps on the day of installation.

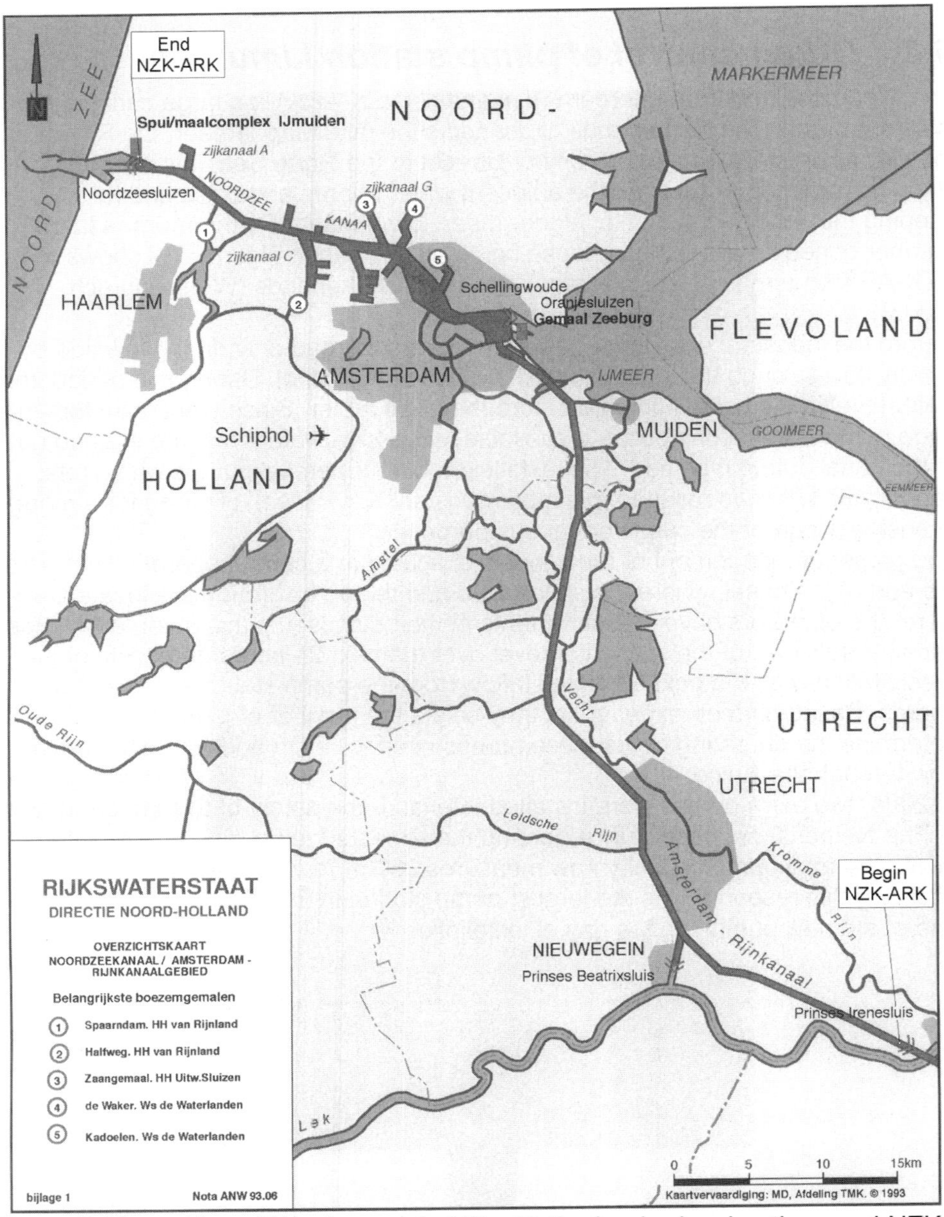

Figure 6.39 Map of Western part of The Netherlands showing the canal NZK-ARK

Figure 6.40 One of the newly installed pumps of 50 m^3/s each

6.3.1 Objective of off/on control of pump station IJmuiden

The energy consumption of each pump is different and depends mainly on the head between canal water level and sea water level. The total energy consumption of this installation costs approximately €700000 per year (Weissenbruch (2003)). By operating the pumps during periods when their energy consumption is low, savings up to 10% can be achieved (Weissenbruch et al. (2004)). These periods occur just before or just after a gravity discharge period of the sluice, as the head is still low. With the newly installed pumps, the selection of the pumps and the operating period becomes more complex. To support the operators in this task, a Decision Support System has been built that advises on the pump usage over the next 12 hours (in The Netherlands this equals approximately one tidal period). The controller is based on Model Predictive Control that minimizes the energy consumption against the canal water level deviation from setpoint:

$$\min_{\vec{P}} J = \sum_{i=0}^{n}\left\{e(k+i\,|\,k)^{T}\cdot Q_{e}\cdot e(k+i\,|\,k)\right\}+\sum_{i=0}^{n-1}\left\{\vec{P}_{e}(k+i\,|\,k)^{T}\cdot \overline{R}_{P_{e}}\cdot \vec{P}_{e}(k+i\,|\,k)\right\}$$

Formula 6.23

where e represents the deviation between water level and setpoint (m), Q_e the penalty on this deviation, $\vec{P}_e$ the vector with the power usage of all pumps (kWatt) and $\overline{R}_{P_e}$ the matrix with the penalties on the power usage of all pumps.

6.3.2 Characteristics of off/on control of pump station IJmuiden

The wide canal is considered to be one large reservoir. The water level in this reservoir is controlled by the sluice gate and the six pumps. The six pumps have a total of ten different stages with different pump capacity and energy consumption. Pump 1 and 3 only have an off or on stage, while pump 2 and 4 have an off, low or high capacity stage. The new pumps are frequency driven, but in the Decision Support System they can only be set to off, 40 m3/s or 50 m3/s. So the solution calculated by the controller must be discrete, such that these stages of the pumps are either off or on. Of course, stages of one pump that exclude each another may not be selected at the same time. If, for example, the stage low capacity of pump 3 is turned on at a certain time step, the stage high capacity of pump 3 must be turned off. Ultimately, the Model Predictive Controller selects the pumps and the stage at which these pumps have to run in such a way that the water level deviations remain small and the smallest amount of energy is consumed.
The sluice flow and the ten stages of the pumps are modeled as separate inputs to the water system. The disturbance flow consists of the various sources of inflow along the canal from direct run-off and inflow from the pump stations of the water boards. The state space model of the canal is:

$$
\left[e(k+1)\right]=\left[1\right]\cdot\left[e(k)\right]+\begin{bmatrix}
-\frac{T_c}{A_s} \\
-\frac{T_c}{A_s}\cdot\frac{Q_{c,1,\max}}{P_{e,1,\max}} \\
-\frac{T_c}{A_s}\cdot\frac{Q_{c,2-low,\max}}{P_{e,2-low,\max}} \\
-\frac{T_c}{A_s}\cdot\frac{Q_{c,2-high,\max}}{P_{e,2-high,\max}} \\
-\frac{T_c}{A_s}\cdot\frac{Q_{c,3,\max}}{P_{e,3,\max}} \\
-\frac{T_c}{A_s}\cdot\frac{Q_{c,4-low,\max}}{P_{e,4-low,\max}} \\
-\frac{T_c}{A_s}\cdot\frac{Q_{c,4-high,\max}}{P_{e,4-high,\max}} \\
-\frac{T_c}{A_s}\cdot\frac{Q_{c,5-40,\max}}{P_{e,5-40,\max}} \\
-\frac{T_c}{A_s}\cdot\frac{Q_{c5-50,\max}}{P_{e,5,-0,\max}} \\
-\frac{T_c}{A_s}\cdot\frac{Q_{c,6-40,\max}}{P_{e,6-40,\max}} \\
-\frac{T_c}{A_s}\cdot\frac{Q_{c,6-50,\max}}{P_{e,6-50,\max}}
\end{bmatrix}^T\cdot\begin{bmatrix}
Q_{c,sluice}(k) \\
P_{e,1}(k) \\
P_{e,2-low}(k) \\
P_{e,2-high}(k) \\
P_{e,3}(k) \\
P_{e,4-low}(k) \\
P_{e,4-high}(k) \\
P_{e,5-40}(k) \\
P_{e,5-50}(k) \\
P_{e,6-40}(k) \\
P_{e,6-50}(k)
\end{bmatrix}+\left[\frac{T_c}{A_s}\right]\cdot\left[Q_d(k)\right]
$$

Formula 6.24

where k represents the time step index, e the deviation between canal water level and setpoint (m), T_c the control time step (s), A_s the storage area of the canal (m^2), $Q_{c,sluice}$ the control flow of the sluice (m^3/s), Q_d the disturbance flow consisting of the summed inflows along the canal (m^3/s) and P_e the power usage of a pump (kWatt). Note that the input matrix is given in transposed form. Initially, the assumption is made that the flow of a pump is proportional to the power by a factor $Q_{c,max}/P_{e,max}$. This assumption is justified when the pump stage is either off or on.

To accurately predict the influence of the tide on the solution, a control time step of 10 minutes is used. With this time step the tidal sea water level consisting of a period of 12.25 hours can be described with sufficient accuracy.
Pumps 1 and 3, 2 and 4 and 5 and 6 are identical in twos. If only one pump of each of these pairs needs to be used, the optimization will always select the first one. To avoid wear and tear of this one pump, an extra weight is used in the optimization that enables the use of a priority of one pump over the other identical one. Ultimately, the objective function becomes:

$$\min_{Q_{c,sluice},\vec{\mathbf{P}}} J =$$

$$\sum_{i=0}^{n}\left\{e(k+i|k)^T \cdot Q_e \cdot e(k+i|k)\right\}+$$

$$\sum_{i=0}^{n-1}\left\{\begin{array}{l}
Q_{c,sluice}(k+i|k)\cdot R_{Q_{c,sluice}}\cdot Q_{c,sluice}(k+i|k)+\\
P_{e,1}(k+i|k)^T\cdot R_P\cdot R_{pr,1}\cdot P_{e,1}(k+i|k)+\\
P_{e,2-low}(k+i|k)^T\cdot R_P\cdot R_{pr,2}\cdot P_{e,2-low}(k+i|k)+\\
P_{e,2-high}(k+i|k)^T\cdot R_P\cdot R_{pr,2}\cdot P_{e,2-high}(k+i|k)+\\
P_{e,3}(k+i|k)^T\cdot R_P\cdot R_{pr,3}\cdot P_{e,3}(k+i|k)+\\
P_{e,4-low}(k+i|k)^T\cdot R_P\cdot R_{pr,4}\cdot P_{e,4-low}(k+i|k)+\\
P_{e,4-high}(k+i|k)^T\cdot R_P\cdot R_{pr,4}\cdot P_{e,4-high}(k+i|k)+\\
P_{e,5-40}(k+i|k)^T\cdot R_P\cdot R_{pr,5}\cdot P_{e,5-40}(k+i|k)+\\
P_{e,5-50}(k+i|k)^T\cdot R_P\cdot R_{pr,5}\cdot P_{e,5-50}(k+i|k)+\\
P_{6-40}(k+i|k)^T\cdot R_P\cdot R_{pr,6}\cdot P_{6-40}(k+i|k)+\\
P_{6-50}(k+i|k)^T\cdot R_P\cdot R_{pr,6}\cdot P_{6-50}(k+i|k)
\end{array}\right\}$$

Formula 6.25

where *e* represents deviation between water level and setpoint (m), Q_e the penalty on this deviation, P_e the power usage of a pump stages (kWatt), R_{Pe} the penalty on the power usage of all pumps. R_{pr} represents the extra weight to get a priority of usage of one pump over the other identical pump. An extra weight slightly lower than 1 means a slightly smaller penalty for this specific pump.

Consequently, this pump will be selected first over the identical pump with an extra weight slightly higher than 1. In the tests, the priority ranking is:

- Pump 1 has priority over identical pump 3;
- Pump 4 has priority over identical pump 2 and;
- Pump 5 has priority over identical pump 6.

The parameter values for the penalties are determined by the method described in Paragraph 3.3, without extensive tuning. The Maximum Allowed Value Estimate of the water level deviation and the power are 0.1 m and 800 kWatt, respectively. The value for $R_{Qc,sluice}$ is derived from the assumption that a water level deviation of 1 millimeter above the setpoint, requires the sluices gate to go to the maximum flow of 500 m^3/s (fully open). The resulting low value of the weight factor $R_{Qc,sluice}$ will cause the gravity driven sluice gate, with a negligible energy consumption, to be selected first when water needs to be discharged. Table 6.5 gives the parameter values used in the objective function of the Model Predictive Controller.

Table 6.5 Parameters of Model Predictive Control IJmuiden

Parameter	Value	Unit
n	73	Steps (12 h)
Q_e	100	-
$R_{Qc,sluice}$	4.0e-10	-
R_{Pe}	1.56e-6	-
$R_{pr,1}$	0.99	-
$R_{pr,2}$	1.01	-
$R_{pr,3}$	1.01	-
$R_{pr,4}$	0.99	-
$R_{pr,5}$	0.99	-
$R_{pr,6}$	1.01	-

6.3.3 Constraints of off/on control of pump station IJmuiden

The demands of the multiple users of the canal require a tight control of the water level in the canal. The water level must remain close to the setpoint of -0.40 meter Mean Sea Level.
The maximum flow of the sluice gate and the maximum power of the pump stages vary in time as a function of the head between canal water level and sea water level. Figure 6.41, 6.42 and 6.43 show the maximum sluice gate flow, maximum pump capacities and maximum power usage, respectively, as function of the head. The sluice gate may start to discharge when the head is larger than 12 centimeter. This is to avoid intrusion of saline sea water. The original pumps can only discharge water when the head is negative, while the newly installed pumps can pump until a positive head of 20 centimeters:

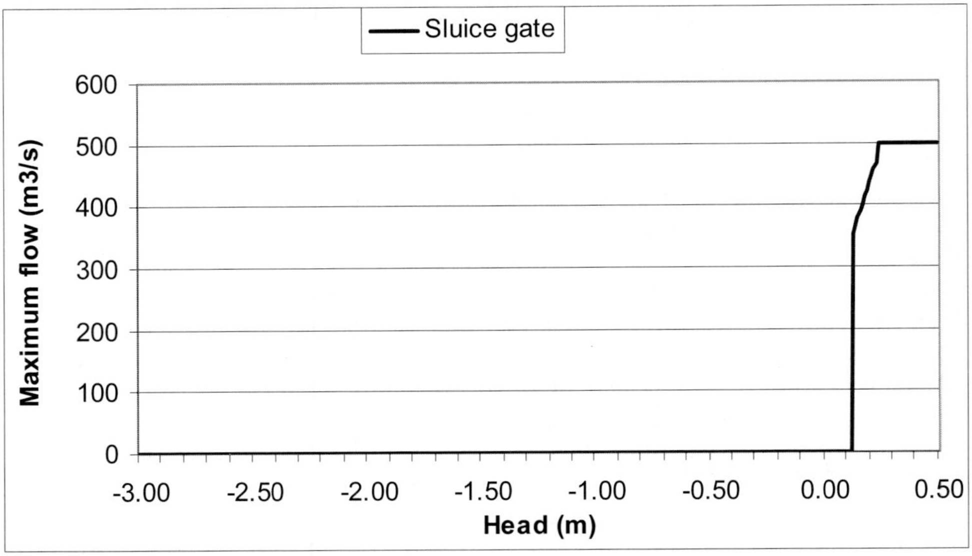

Figure 6.41 Maximum sluice gate flow as function of head between canal and sea

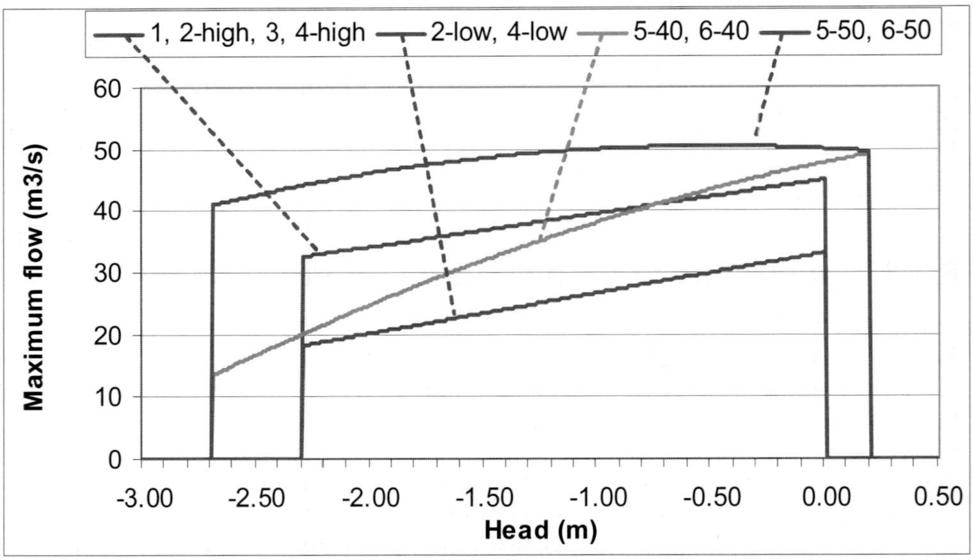

Figure 6.42 Maximum pump capacity of pump stages as function of head between canal and sea

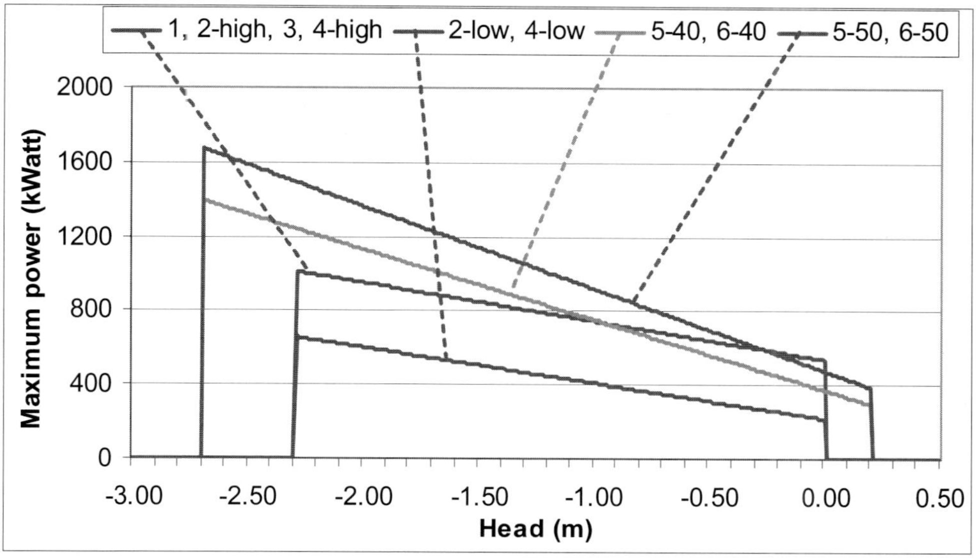

Figure 6.43 Maximum power usage of pump stages as function of head between canal and sea

Within the computation of one solution, the maximum input constraints are calculated in advance based on the canal water level of the previous control time step. In reality this maximum flow and maximum power will change with the change in water level during a computation. To improve this explicit implementation, the computation is repeated a number of times in an iterative loop. In each computation, the maximum flow through the sluice gate and the maximum power of each pump is computed by weighing the water level solution of the previous time step and the new water level solution. The weight factor is set at 0.5 similar to the Crank Nicolson method (Stelling & Booij (1994)). Sequential configuration of Model Predictive Control, as described in Chapter 4, allows for this iterative improvement of the accuracy of the maximum constraints within the optimization.

The sequential configuration of MPC needs to result in an off/on switching pattern of the different pump stages. A single optimization results in a continuous solution and needs to be converted into a discrete solution. This is done by step-wise rounding. After the first optimization, the continuous solutions of all pump flows over the prediction horizon are rounded to either off or on, when the difference between the continuous solution and off/on solution is less than a certain margin. In the next optimization, the pump flows that are rounded are fixed, reducing the degrees of freedom for the solution. At the time steps in the prediction horizon where no rounding has taken place, the optimization will compensate for the rounding at other time steps in order to keep the objective function as low as possible. In this way, the solution will evolve into a more off/on shaped solution. In the rounding of the second iteration, the margin for rounding is increased. This

sequential optimization and rounding is repeated until the margin becomes half the difference between zero flow and maximum flow. In this last iteration, all flows that are not fixed yet will consequently be rounded to either off or on. Figure 4.3 shows an example of the step-wise rounding of one of the pump flows based on three sequential iteration steps.
An additional constraint that needs to be incorporated in the controller is that the pumps may not be turned on and off every control time step. Because of the large size of the pumps, it takes time and extra energy before the inertia of the rotating parts is overcome. The time before a pump may switch again, is 30 minutes, which is equal to 3 control time steps. This is incorporated into the sequential configured Model Predictive Controller as an extra condition for the rounding. Here, a loop from the present time step to the end of the prediction horizon *n* is executed, which checks if three sequel steps have the same rounded value. Figure 6.44 shows the decision diagram that is used in each iteration step.

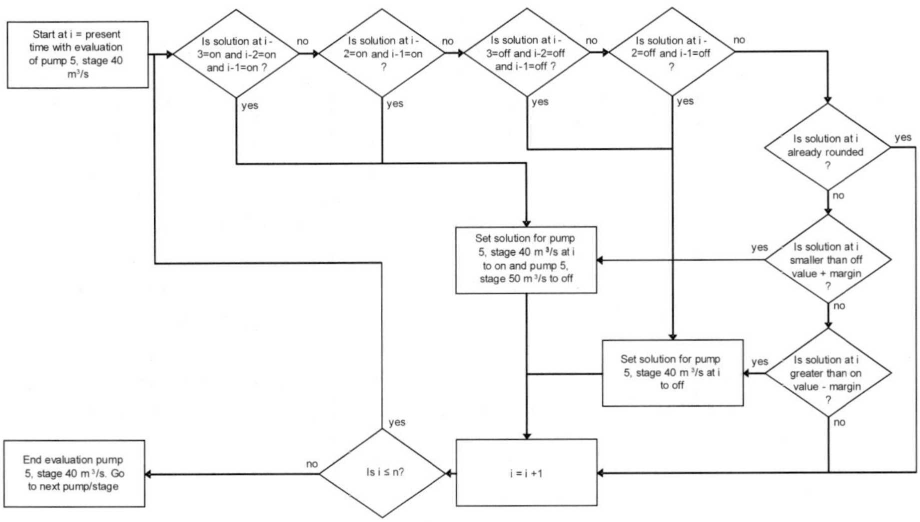

Figure 6.44 Decision diagram to ensure pump switching only each 30 minutes

Summarizing, the constraints are:

- The sluice gate flow must be equal to or greater than zero and equal or lower than the maximum sluice gate flow given the water level head;
- The pumped flows must be equal to or greater than zero and equal or smaller than the maximum capacity given the water level head;
- The power usage must be equal to or greater than zero and equal or smaller than the maximum power given the water level head;
- The pump stages must be either off or on;

- The pump stages must remain in the same status (off or on) for at least 30 minutes;
- If one stage of a pump is rounded to an on status, the other stage of the same pump must be fixed to the off state (as they exclude one another) and vice versa.

6.3.4 Results of off/on control of pump station IJmuiden

Since 2005, the Model Predictive Controller as previously described, is available in the Decision Support System at the central operating room of pump station IJmuiden (Weissenbruch et al. (2004), Overloop et al. (2005c)). It can run with only one iteration step as straight forward Model Predictive Controller and as sequential configuration of Model Predictive Control with three iteration steps. The results given below show the difference between both configurations as tested on a hydro-dynamic model (Sobek (2000)) of the canal with the modeled pump station and sluice gate. The hydro-dynamic model is linked to Matlab (MathWorks (1992)). Matlab calls the actual controller, as programmed and compiled in C-code (Visual C++ (2000)). This setup was made to enable debugging of the actual controller in a safe environment, before the controller is implemented in the actual DSS in the control room. The test is run based on a heavy storm event, so that the use of only the sluice gate is not sufficient to keep the water level below the maximum allowed water level (approximately -0.30 mMSL). Instead, the controller must select pump stages that discharge water at the lowest possible energy cost. The open loop discharges of sluice and pump stages and the water levels are presented in Figures 6.45 and 6.46. This solution is the result of the first optimization over 12 hours at the beginning of the simulation. Figures 6.47 and 6.48 present the results of the closed loop solution over these 12 hours of the hydro-dynamic model. As sequential configuration of Model Predictive Control requires considerable computation time, this controller is run once every hour. As the control time step over the prediction horizon is 10 minutes, each hour, six control actions are put into effect.

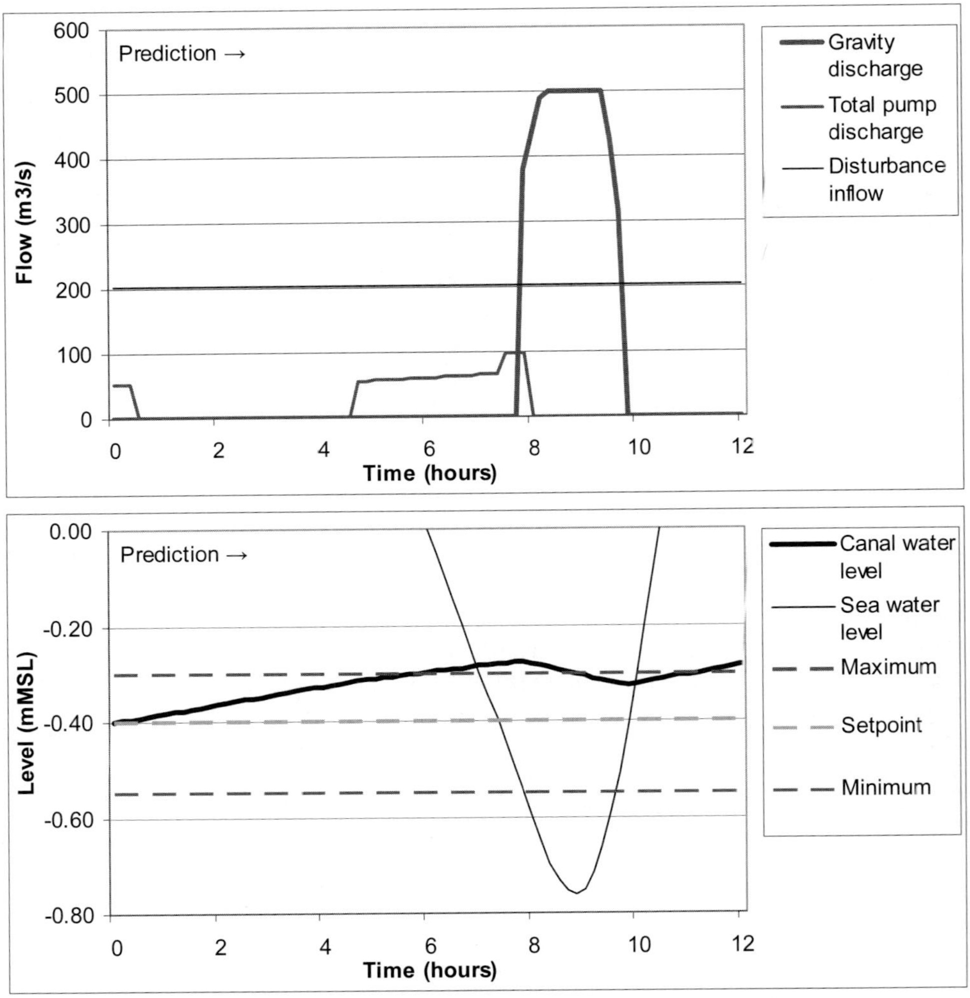

Figure 6.45 Results of open loop solution of MPC with 1 iteration step

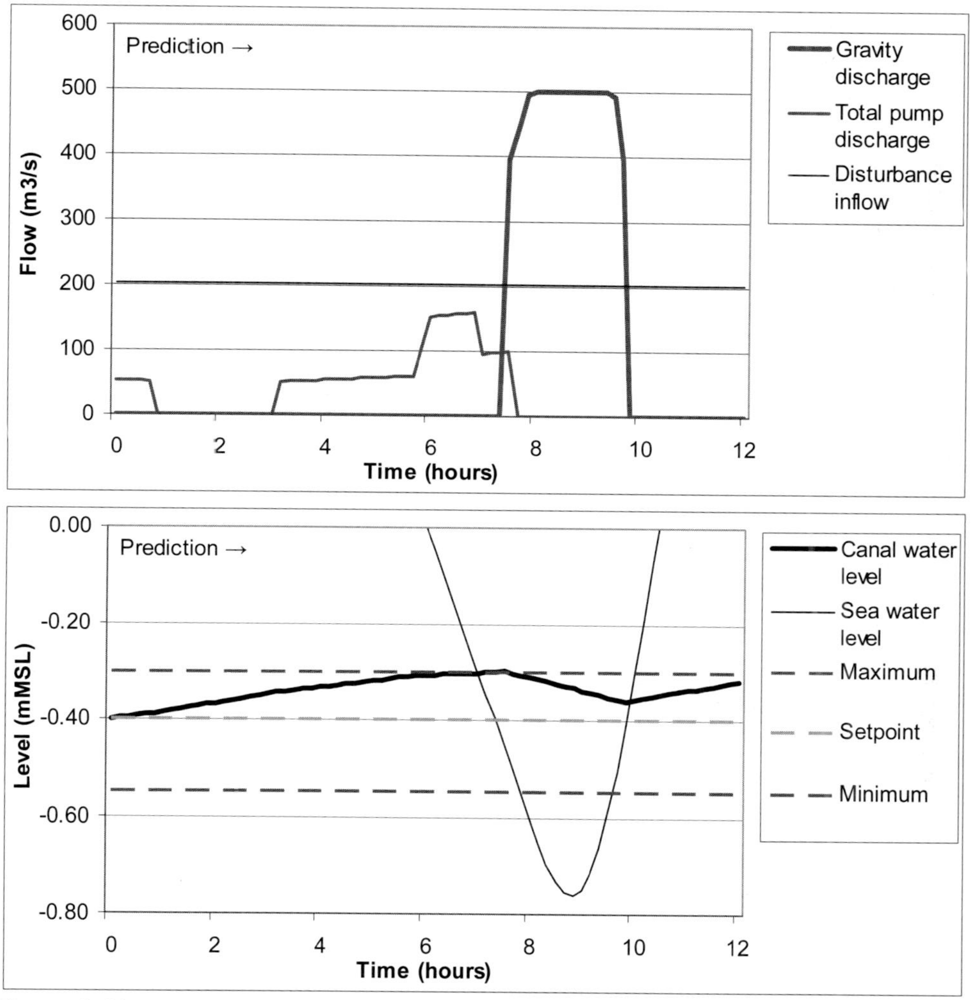

Figure 6.46 Results of open loop solution of scMPC with 3 iteration steps

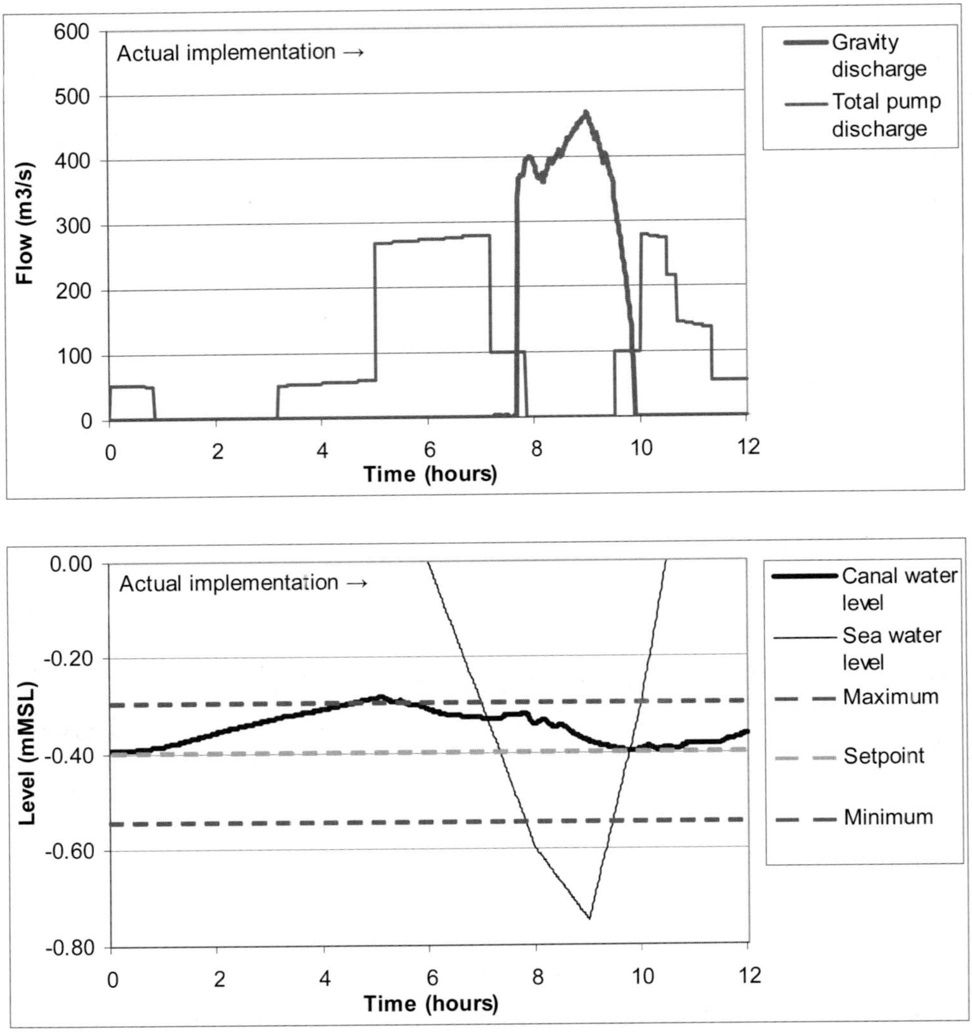

Figure 6.47 Results of closed loop solution of MPC with 1 iteration step

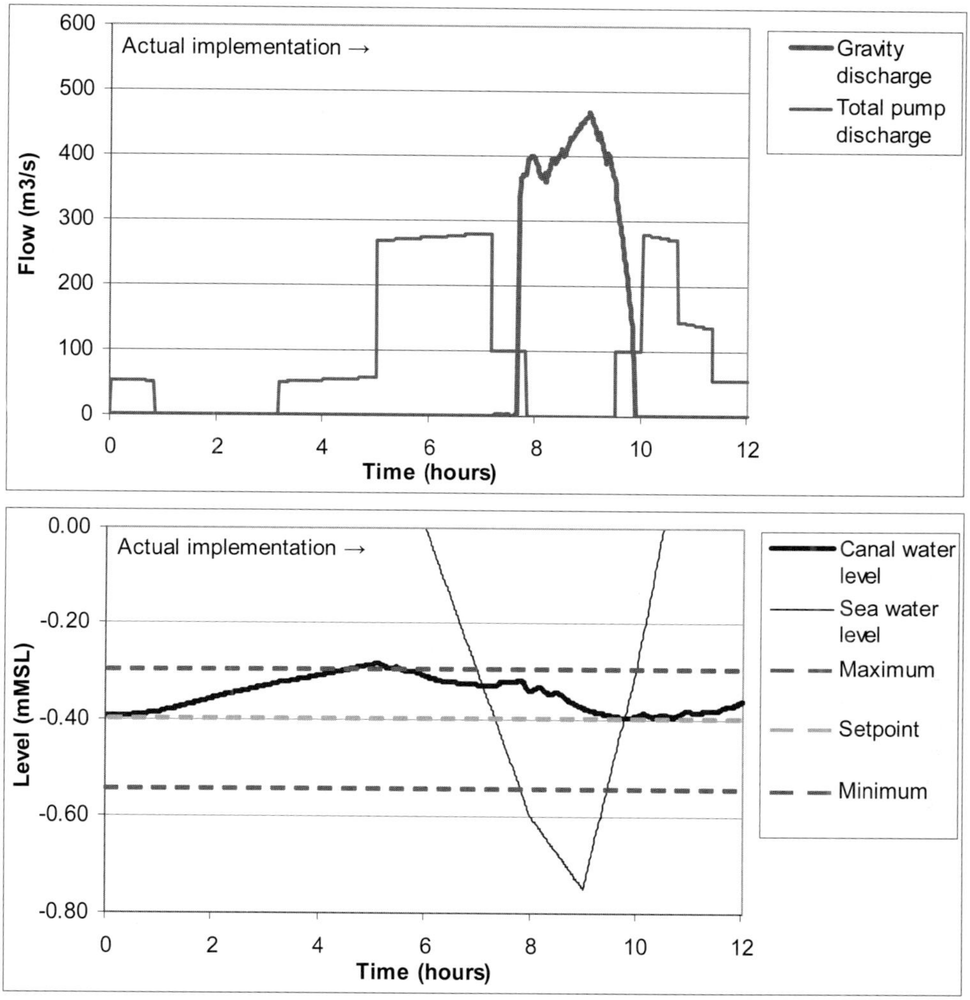

Figure 6.48 Results of closed loop solution of scMPC with 3 iteration steps

To show that the constraints are met and that the priority ranking of one pump over the identical other pump functions well, the operation of each pump stage of the closed loop sequential configuration of MPC solution is shown in Figure 6.49.

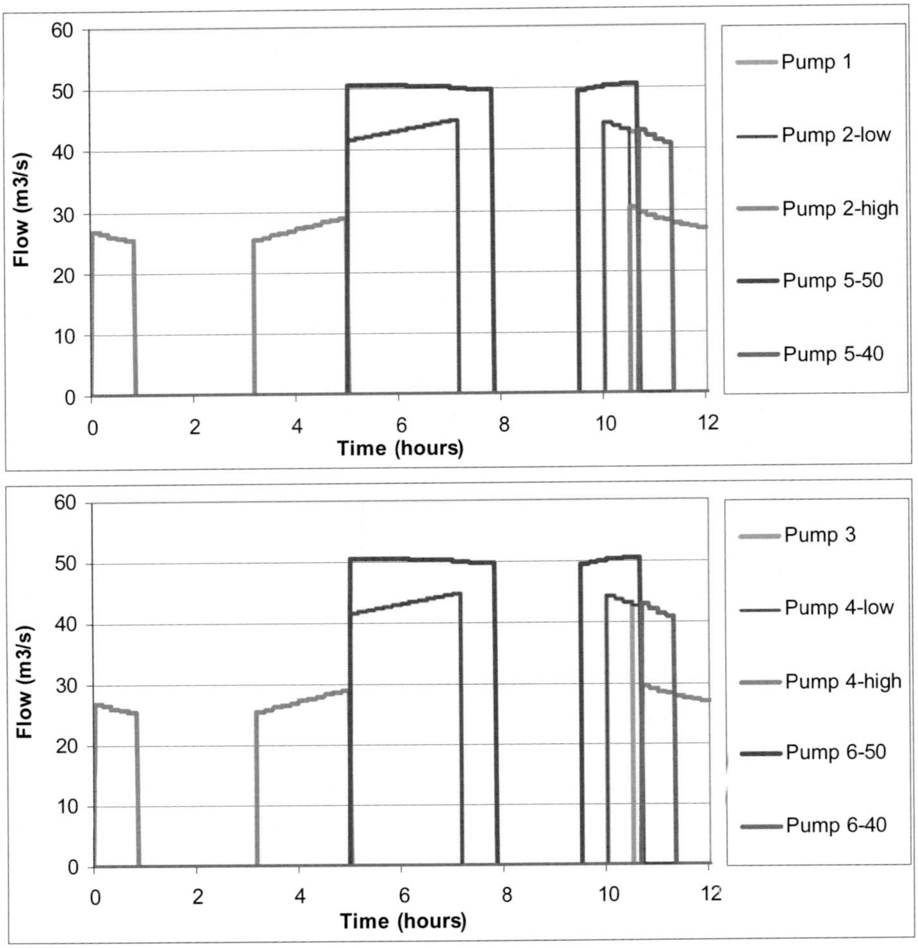

Figure 6.49 Operation of pump stages

6.3.5 Discussion on results of off/on control of pump station IJmuiden

The results show that the controller is able to generate open loop and closed loop solutions in which all constraints are met. The inputs are all higher or equal to zero and smaller or equal to the maximum value. The switching between the different stages of the pumps occurs every 30 minutes or less often and stages that exclude one another are not running at the same time. The priority ranking of identical pumps functions appropriately. This can be seen in Figure 6.49 at 10:45. At that moment pump 3 switches off, while pump 1 with a higher priority remains on for 10 more minutes. At the same time, pump 2 switches off, while pump 4 with a higher priority remains on for another 10 minutes.

The open loop solution of sequential configuration of Model Predictive Control as given in Figure 6.46, in which the solution is calculated based on three iteration steps, remains within the allowed range around setpoint, while Model Predictive Control based on only one iteration step as shown in Figure 6.45, violates the constraints. To achieve this, it deploys more pump stages. This is revealed by comparing the flows in Figure 6.45 and Figure 6.46.
The closed loop solutions show a higher utilization of the pumps compared to the open loop solutions. This is caused by the prediction horizon of 12 hours. The prediction horizon ends just after a low tide, with an open loop water level that is close to the maximum water level. The Model Predictive Controller is not able to look further into the future than this moment, so it will not anticipate the part of the high tide that occurs just after the end of the prediction horizon. In the closed loop solutions, the prediction horizon has moved forward in time, anticipating the rise in water level due to this high tide. To create extra storage for the limited outflow due to the high tide, the resulting water levels at the end of the closed loop solutions are lowered at the cost of higher energy consumption. An additional reason for the pump flows being higher in the closed loop situation is that the actual gravity discharge through the sluice gate is smaller than assumed in the internal model of the controller.
The close loop solution of sequential configured MPC performs slightly better than MPC when the water level deviation from setpoint is considered. This is achieved at the cost of a small amount of extra energy consumption. To make a fair comparison to decide which controller performance better, some characteristics of the results are summarized in Table 6.6:

Table 6.6 Performance characteristics of controller configurations

Controller configu-ration	Open Loop or Closed Loop	Gravity volume over 12 hours (m^3)	Pump volume over 12 hours (m^3)	Energy consumption over 12 hours (kWh)	Objective function over 12 hours	Water level after 12 hours (mMSL)
MPC	OL	3364383	892563	2263.037	56.078	-0.284
scMPC	OL	4069902	1555194	4771.310	37.834	-0.319
MPC	CL	2980716	4168305	15002.679	104.479	-0.365
scMPC	CL	2980703	4201218	15164.104	105.538	-0.368

The closed loop solutions of MPC and sequential configuration of MPC show a great resemblance. The energy consumption is slightly higher for sequential configured MPC and so is the objective function value. The final water level resulting from sequential configuration of MPC is slightly lower. The reason for the sequential controller not outperforming MPC can be found in the mismatch between the actual system that is modeled in great detail, while the internal model uses only one reservoir for the entire canal. The total disturbance flow discharges in this reservoir, while in the actual system the disturbance flow is distributed over the entire length of the canal. Further improvements of the internal model are required. However, the potential of using sequential configuration of MPC with sequential rounding over MPC, is very clear from the open loop tests. Sequential configured MPC in open loop is able to discharge more flow through the sluice gate. Even though the pump flow and consequently the energy consumption are higher, the objective function value is much lower. This is due to the final water level being considerably lower. Even more importantly, the maximum water level remains within the allowed range around setpoint when sequential configuration of MPC with sequential rounding is applied (see Figure 6.46).

6.4 *Stochastic control of drainage system Delfland*

In Paragraph 6.1, various controllers are tested on the drainage system Delfland. It showed that controllers that use predictions improve the performance of the controlled water system. This is in concordance with the way the operators usually manage the water system. They use predictions to take control actions in advance. An additional fact is that the predictions they use are, by definition, inaccurate to some extent. These predictions result from models that can never perfectly describe reality.
In this paragraph, the results of multiple model configuration of Model Predictive Control applied to the drainage system of Delfland are presented. Multiple model configuration of MPC is a method to incorporate the uncertainty of prediction models into the control actions by using multiple models of the water system in parallel. It is based on the risk-approach in which risk is defined as the probability of occurrence multiplied by the resulting costs.

6.4.1 Objective of stochastic control of drainage system Delfland

The objectives of stochastic control of the drainage system Delfland are the same as described in Paragraph 6.1.1. Only now the assumption is made that the disturbance model describing the rainfall-runoff processes, is uncertain. The precipitation is probably even more uncertain, but here a different element in the control loop is taken into account for the following reason:

- As this research is carried out within the field of water management, knowledge of the validity of the hydrologic disturbance model is more readily available. Uncertainties in the precipitation should be researched by meteorologists;
- The presented method can be generally applied to all elements of the control loop. This research only serves to study the general applicability of multiple model configuration of MPC;
- In the Model Predictive Controller implemented at the central operating room at the water board of Delfland, the forecast of precipitation that is received from the meteorological institute does not include probabilities (yet).

Therefore, in addition to the before mentioned objectives, the water levels of all multiple models that result from disturbances with a different probability must remain within the range around setpoint.

6.4.2 Characteristics of stochastic control of drainage system Delfland

The characteristics of the water system of Delfland are described in Paragraph 6.1.2. Now the disturbance flow as given in Figure 6.6 is considered to be

uncertain. To what extent the disturbance flow is uncertain is estimated by Monte Carlo analysis (Sobel' (1994), Manno (1999)). Here, a number of simulations of the disturbance model are repeated in which the parameters of the disturbance model are varied. This results in a set of solutions in which each solution has the same probability. If the number of repetitions approaches infinity, the actual solution is expected to lie inside the outer uncertainty bounds. This is based on the assumption that only the parameters are considered uncertain and that the varied parameter values that are used represent the range of actual parameter values.
In this research, each parameter value is selected from a Normal distribution. The mean value and the standard deviation of the Normal distributions of each parameter are found by expert judgment of both hydrologists and operators of the water board of Delfland. The number of repetitions is set at 50 to keep the execution workable. As each simulation is independent, this number can be set much higher by using parallel processing. Figure 6.50 presents all solutions of the predicted disturbance flow over 24 hours at the first time step of the simulation of September 13th and 14th 1998.

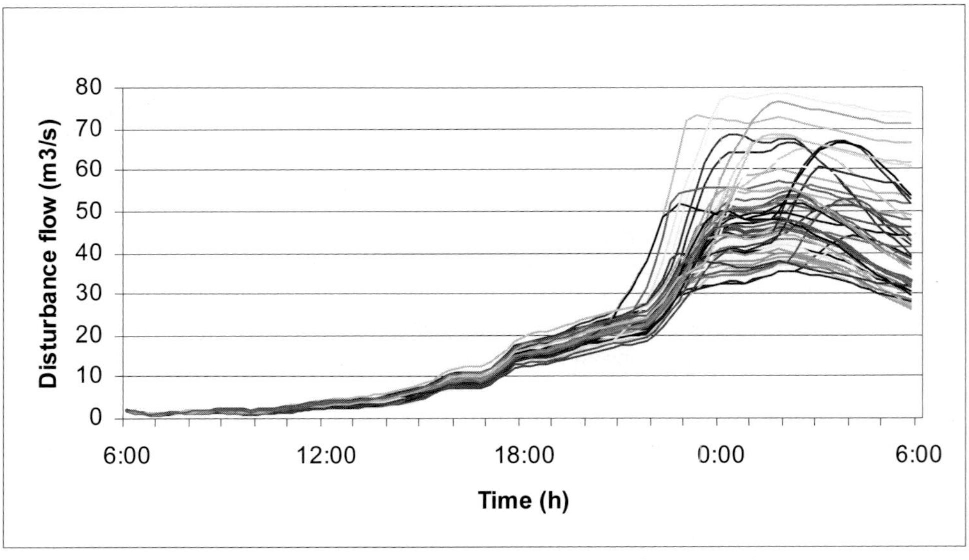

Figure 6.50 Fifty solutions of disturbance model with uncertain parameters

Next, the disturbance flows are sorted by increasing the total volume over 24 hours. At this point, the assumption is made that the higher the total amount of volume, the higher the peak water levels will be. For a reservoir model, this seems a logical assumption, but the validity of this assumption is not studied in this research. A final step is to average a total set of solutions and estimate the probability by summing the probability of occurrence of each member in the set. In this way, a minimum, average and maximum scenario model is estimated. In this test, the minimum, average and maximum scenarios are the lowest 5, middle 40 and highest 5 solutions, respectively. The probabilities P_{min}, P_{avg} and P_{max} of these scenarios are 10%, 80% and 10%, respectively. Figure 6.51 shows the model of the three scenarios, while Figure 6.52 gives the open loop disturbance flows.

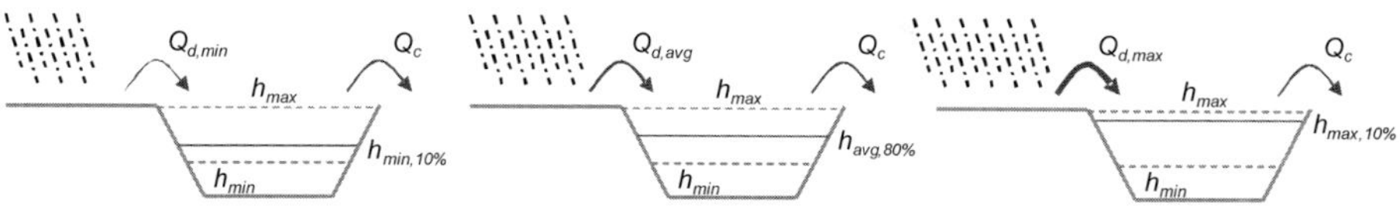

Figure 6.51 Multiple models of three scenarios

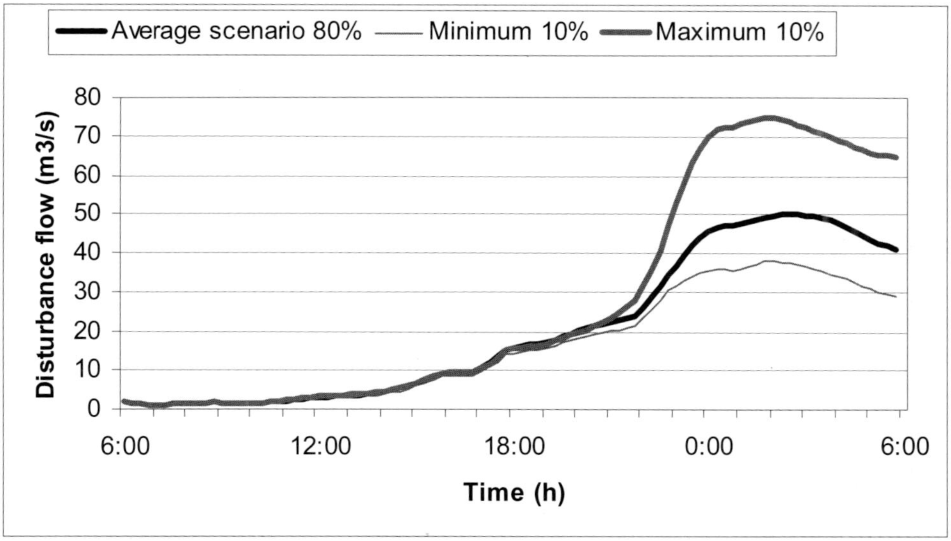

Figure 6.52 Three scenarios of disturbance model with uncertain parameters

An operator confronted with a prediction as shown in Figure 6.52 will base his control actions mainly on the average scenario as it has the highest probability. Only if the minimum or maximum scenario results in high cost, the operator will weight this scenario more in his decision making. In fact, the operator is combining probability and cost into a risk-approach. Risk is defined as probability multiplied by the resulting costs. In water systems, costs are directly related to the water level. High water levels result in damage due to inundation. Low water levels result in damage due to dike instabilities or claims by house boat owners. Based on the objective function as given in Formula 6.9, costs can be identified as the square of the water level deviation from setpoint. This allows for incorporation of risk in a new objective function by multiplying all variables related to the square of the water level deviation by the probability. In Formula 6.26, this new objective function is given in comparison to the simpler objective function as presented in Formula 6.9:

$$
\begin{aligned}
&\min_{\Delta Q_c, u^*_{avg}, u^*_{\min}, u^*_{\max}} J = \\
&\sum_{i=0}^{n} \left\{ \begin{array}{l}
P_{avg} \cdot e_{avg}(k+i|k)^T \cdot Q_e \cdot e_{avg}(k+i|k) + \\
P_{avg} \cdot \Delta e_{avg}(k+i|k)^T \cdot Q_{\Delta e} \cdot \Delta e_{avg}(k+i|k) + \\
P_{avg} \cdot e^*_{avg}(k+i|k)^T \cdot Q_{e^*} \cdot e^*_{avg}(k+i|k) + \\
P_{avg} \cdot u^*_{avg}(k+i|k)^T \cdot R_{u^*} \cdot u^*_{avg}(k+i|k) + \\
P_{\min} \cdot e_{\min}(k+i|k)^T \cdot Q_e \cdot e_{\min}(k+i|k) + \\
P_{\min} \cdot \Delta e_{\min}(k+i|k)^T \cdot Q_{\Delta e} \cdot \Delta e_{\min}(k+i|k) + \\
P_{\min} \cdot e^*_{\min}(k+i|k)^T \cdot Q_{e^*} \cdot e^*_{\min}(k+i|k) + \\
P_{\min} \cdot u^*_{\min}(k+i|k)^T \cdot R_{u^*} \cdot u^*_{\min}(k+i|k) + \\
P_{\max} \cdot e_{\max}(k+i|k)^T \cdot Q_e \cdot e_{\max}(k+i|k) + \\
P_{\max} \cdot \Delta e_{\max}(k+i|k)^T \cdot Q_{\Delta e} \cdot \Delta e_{\max}(k+i|k) + \\
P_{avg} \cdot e^*_{\max}(k+i|k)^T \cdot Q_{e^*} \cdot e^*_{\max}(k+i|k) + \\
P_{\max} \cdot u^*_{\max}(k+i|k)^T \cdot R_{u^*} \cdot u^*_{\max}(k+i|k)
\end{array} \right\} + \\
&\sum_{i=0}^{n-1} \left\{ \Delta Q_c(k+i|k)^T \cdot R_{\Delta Q_c} \cdot \Delta Q_c(k+i|k) \right\}
\end{aligned}
$$

Formula 6.26

The consistency of this formula can be checked by assuming identical disturbance flows for the three scenarios (for example if the standard deviations of the Normal distributions of all parameters are set at zero). The resulting water levels of the three scenarios will then be identical. Multiplication by each probability and summation leads to the exact same result as given by Formula 6.9.

6.4.3 Constraints of stochastic control of drainage system Delfland

The same constraints as described in Paragraph 6.1.3 are applied to the three resulting water levels (see Formula 6.6). This means that if the resulting water level of one of the models threatens to violate a constraint, the control flow will be adjusted to avoid this.

6.4.4 Results of stochastic control of drainage system Delfland

For every control step, the predictions of the disturbance flows into the storage canals of the minimum, average and maximum scenarios are used. Figure 6.52 presents the disturbance flows on September 13th 1998 at 6:00 in the morning. This extreme storm event, as described in Paragraph 6.1.4, is tested as it will bring one or more scenarios into the constraints. The actual situation is modeled in which the forecast of the precipitation is underestimated by 50% compared to the actual, measured precipitation (see 6.1.4 for more details). The solution of the optimization of the multiple model configured Model Predictive Controller at that moment given the three disturbance flows, is shown in Figure 6.53. The water levels that result from this open loop solution are given in Figure 6.54.

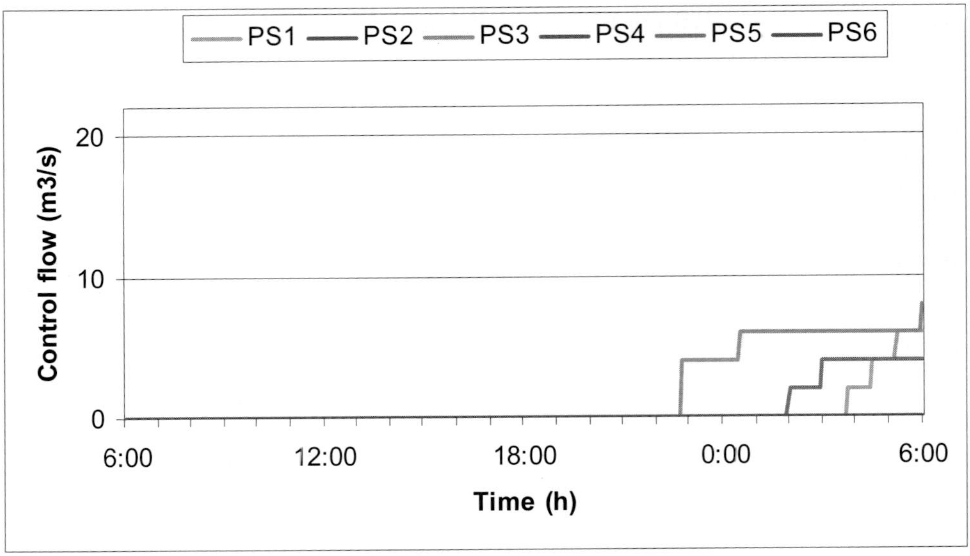

Figure 6.53 Discharge of pump stations when multiple model configuration of Model Predictive Control is applied in open loop (1998)

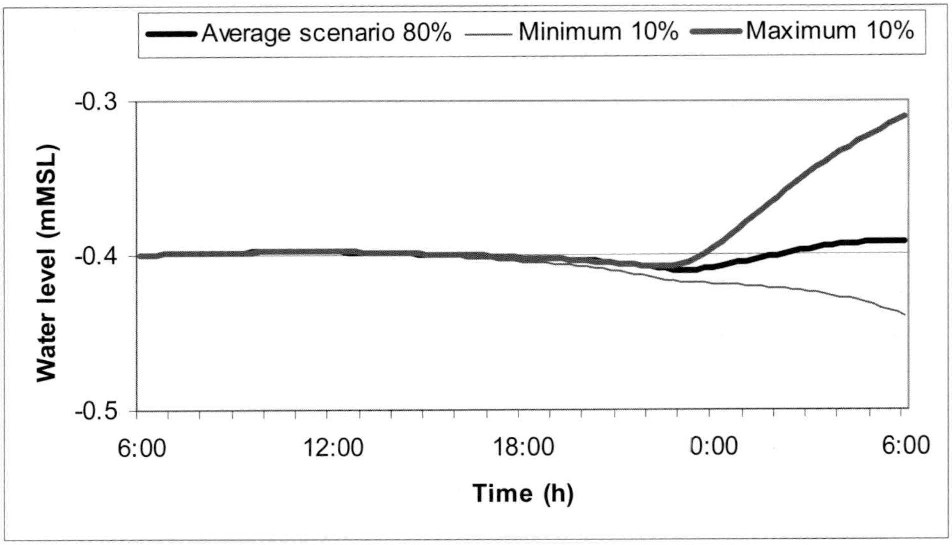

Figure 6.54 Water levels of three scenarios when multiple model configuration of Model Predictive Control is applied in open loop (1998)

In the following Figures 6.55 and 6.56 the closed loop results of the multiple model configured Model Predictive Controller are presented.

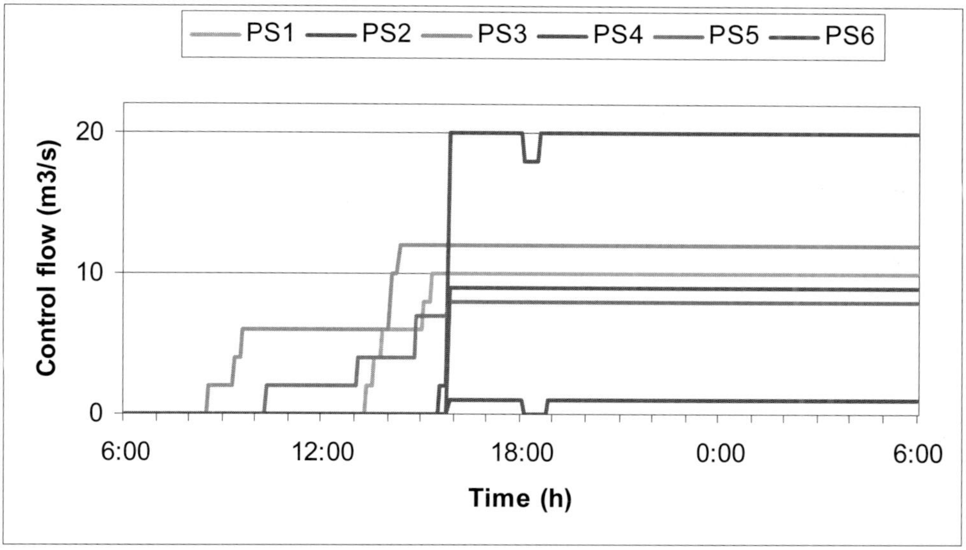

Figure 6.55 Discharge of pump stations when multiple model configuration of Model Predictive Control is applied in closed loop (1998)

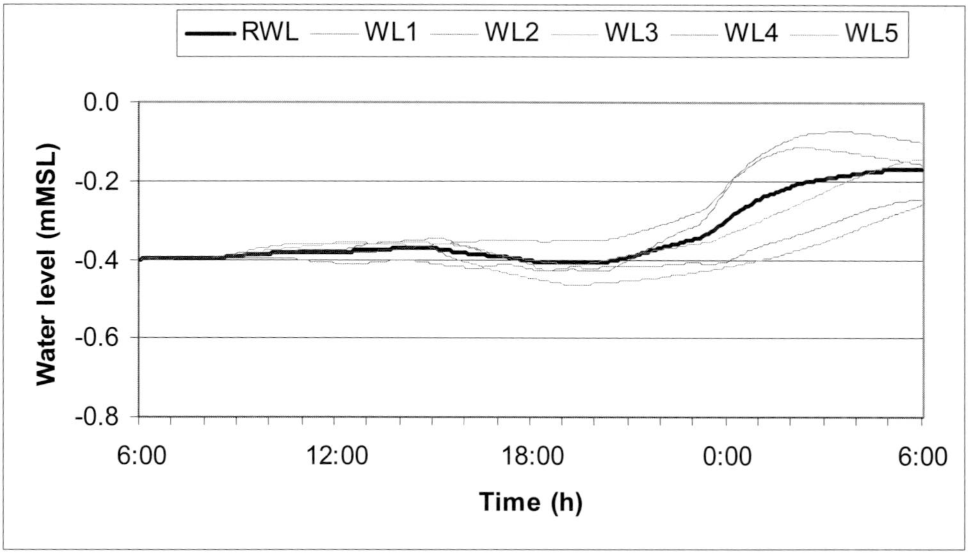

Figure 6.56 Representative water level and distributed water levels when multiple model configuration of Model Predictive Control is applied in closed loop (1998)

6.4.5 Discussion on results of stochastic control of drainage system Delfland

The open loop results given in Figure 6.53 and 6.54 are logical. The control flow is calculated in such a way that the average model with the high probability of occurrence is symmetrical around setpoint. Because of the low probability of the minimum and maximum scenario, the risk of these scenarios is low and the controller does not react strongly to them.

The closed loop results show the influence of the constraints. During the simulation, the water levels and the disturbance flow start to rise, as the forecast of the precipitation is underestimated. The maximum scenario results in an open loop water level that violates the (soft) constraint of the maximum allowed water level sooner than the average scenario. In case the maximum scenario violates the constraint, a very high cost is calculated. This results in a high risk, even though the probability of this maximum model is low. Therefore, the control flow is adjusted earlier than in the case where only the average model is used. This can be seen by comparing Figure 6.55 of multiple model configuration of Model Predictive Control to Figure 6.24 in which the same test is executed with Model Predictive Control based on only one model. The water levels of multiple model configuration of MPC as presented in Figure 6.56 result in a 2 centimeter lower peak compared to MPC as presented in Figure 6.25. Again, it must be stressed that a much higher improvement can be achieved when the forecast of the precipitation is improved. This forecast has a higher uncertainty than the parameters as used in this test based on the uncertain disturbance model.

6.5 *Centralized control of irrigation canal W-M*

The irrigation canal W-M is located in Arizona, South of Phoenix (USA). The canal is part of a larger branched network of irrigation canals managed by the Maricopa-Stanfield Irrigation and Drainage District. Schuurmans & Ellerbeck (1995a) and Malda (2005) give a description of the water system. The canal supplies water to a number of large farms. The length of the canal is almost 10 kilometers and the maximum capacity at the head gate is 2.8 m^3/s. Figure 6.57 shows the location of the irrigation canal.

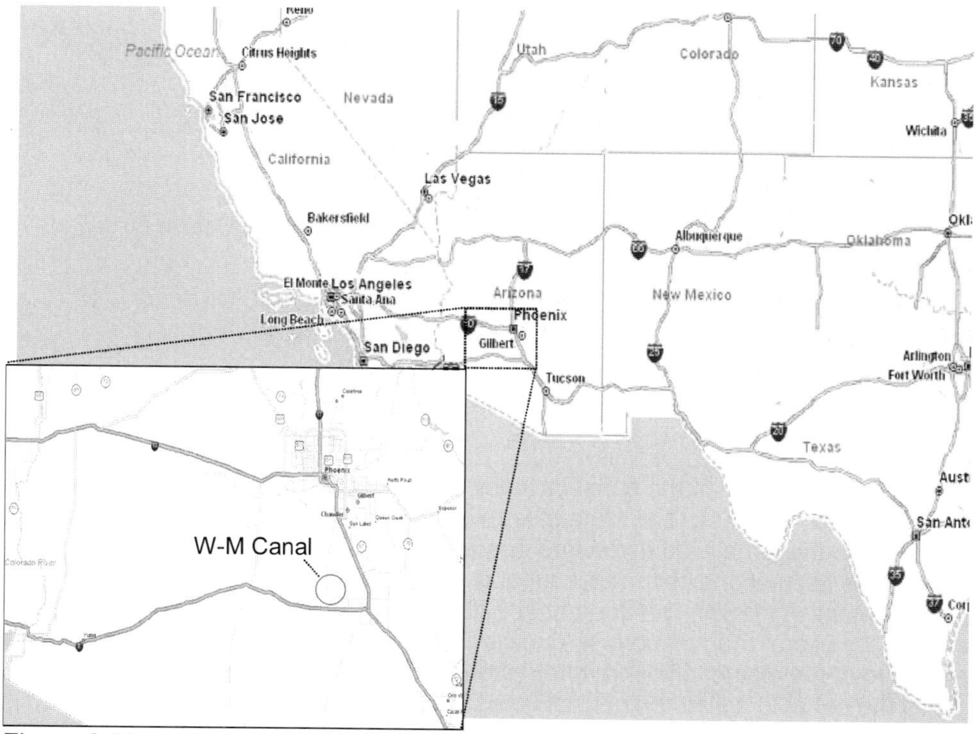

Figure 6.57 Location of W-M Canal

The canal consists of 8 canal reaches in series, although the last canal reach is not in service anymore. The reaches are separated by control structures. Each control structure consists of an adjustable undershot control gate in parallel with weirs with a fixed crest level at both sides of the undershot gate. The offtake undershot gates are located at the end of each canal reach. Behind the offtake gates, the offtake flows are conveyed to the secondary canals by culverts. Figure 6.58 is a picture of one of the control structures.

Figure 6.58 Control structure WM6 in W-M Canal

Originally, the design of the canal included local upstream water level control in the eight canal reaches. The state of knowledge on controlling water systems in 1990, when the canal was taken into operation, proved to be insufficient. Large water level fluctuations occurred, probably due to disturbance amplification (Schuurmans J. (1997), Overloop et al. (2005b)). Since then, management of the canal is by use of manual control. Orders for water from farmers are centrally recorded at the district office from telephone calls. So-called ditch riders drive along the canal to adjust the control gates and offtake gates. The transition from one steady state to another, takes multiple rides up and down the canal, changing gates at all locations. Based on current knowledge, it is possible to control the water levels with automatic controllers. To demonstrate this, the United States Water Conservation Laboratory in Phoenix has tested various types of LQR-controllers in combination with an accurate feedforward controller (Bautista et al. (2003)). These tests show tight control of the water levels (Clemmens et al. (2005)). During the growing season, when the canal has to be operated close to its maximum capacity, the LQR-feedback controller in combination with feedforward might violate constraints on structure flows and water levels. Also, if the maximum number of gate changes have to be restricted, a more advanced controller becomes necessary. Model Predictive Control resembles the LQRcontroller in combination with feedforward, but can also take these constraints into account in a systematic way.

6.5.1 Objective of centralized control of irrigation canal W-M

The objective of the management of the W-M Canal is to deliver the ordered amount of water to the farmers at the right moment, without spilling water. When the offtakes are set at the right time at the correct flow rate, this objective can be translated into keeping the water levels in the canal reaches as close to setpoint as possible. At the same time, this should be done with as few changes to the structures as possible. Under manual execution of the control actions, this results in fewer visits to all sites and in case of automated control gates, it reduces wear and tear of the moving parts. The objective function that implements these specifications is:

$$\min_{\Delta Q_c} J =$$

$$\sum_{i=0}^{n}\sum_{j=1}^{m}\left\{e_j(k+i\,|\,k)\cdot Q_{e,j}\cdot e_j(k+i\,|\,k)\right\}+$$

$$\sum_{i=0}^{n}\sum_{j=1}^{m}\left\{\Delta e_j(k+i\,|\,k)\cdot Q_{\Delta e,j}\cdot \Delta e_j(k+i\,|\,k)\right\}+ \qquad \text{Formula 6.27}$$

$$\sum_{i=0}^{n-1}\sum_{j=1}^{l}\left\{\Delta Q_{c,j}(k+i\,|\,k)\cdot R_{\Delta Q_c,j}\cdot \Delta Q_{c,j}(k+i\,|\,k)\right\}$$

where J represents the objective function that needs to be minimized, n the number of steps over the prediction horizon, m the number of canal reaches, l the number of structures, e_j the water level deviation from setpoint in reach j (m), $Q_{e,j}$ the penalty on this error, Δe_j the change in water level deviation in reach j (m), $Q_{\Delta e,j}$ the penalty on this change in error, $\Delta Q_{c,j}$ the change in control flow of structure j (m^3/s) and $R_{\Delta u}$ the penalty on this change in structure setting. The change in water level deviation and change in control flow are defined as:

$$\Delta e(k) = e(k) - e(k-1) \qquad \text{Formula 6.28}$$

$$\Delta Q_c(k) = Q_c(k) - Q_c(k-1) \qquad \text{Formula 6.29}$$

This selection of variables in the objective function results in control actions that bring the water levels back to the desired levels without static deviation.
The prediction time horizon has to be selected longer than the delay time of the total canal. In this way, the water level response in the last controlled canal reach

to all control actions, even actions taken at the head gate, are included in the optimization.

6.5.2 Characteristics of centralized control of irrigation canal W-M

The first 5000 meters of the canal are steep. Here, super-critical flow occurs. The second part is more level and has larger storage areas. Figure 6.59 shows the longitudinal profile of the canal at low flow steady state, as modeled in an accurate hydro-dynamic model (Sobek (2000)). This model is calibrated by the Manning friction coefficient of the canal reaches and the contraction coefficient of the undershot control gates.

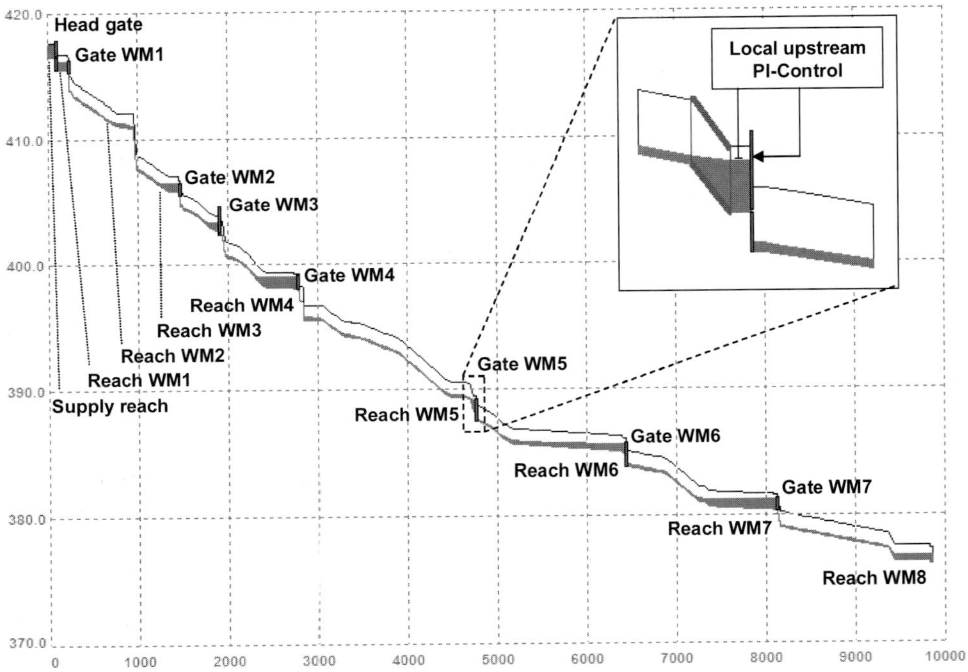

Figure 6.59 Longitudinal profile of modeled W-M Canal

The control structures and offtake structure are free flowing undershot gates that are controlled by flow controllers. This permits the use of control flows in the centralized controller as input to the internal model. The heights of the gates are calculated based on the control flow by using a bi-sectional method (Schuurmans J. (1997)) applied to Formula 2.9. The control time step of this flow controller is 2 minutes.

Directly behind the control structures and at the bottom level drops in the reach canals, culverts are used to convey the water down smoothly. In canal reach WM5 a culvert is constructed under a road directly in front of control gate WM5. This causes a very small back water area of this reach. The values of the storage areas of the back water area of all reaches at low, average and high flow are given in Table 2.3 in Paragraph 2.2.2., to illustrate the variability of this value at the different operating points. The small storage area in canal reach WM5 requires a small control time step. This would result in a very high number of prediction steps given a required prediction time horizon. To avoid the computational burden that comes with this long prediction horizon, canal reach WM5 is controlled by a local upstream PI-controller on gate WM5. Figure 6.59 contains an enlarged illustration of gate WM5.
The estimates of the delay times of the canal reaches at low, average and high flow are also given in Table 2.3 in Paragraph 2.2.2. The prediction time horizon is selected to be 2 hours, as this is higher than the summation of the delay times of all reaches at all flow conditions. Canal reach WM5 is taken out of the centralized controller, as it is locally controlled. However, the delay time of this reach has to be taken into account in the central controller. The local controller has a high performance, so any flow transition is passed almost unchanged. Consequently, canal reach WM5 can be considered as conveyance reach, which only passes water through towards canal reach WM6. In the overall controlled system, canal reaches WM5 and WM6 are considered to be merged into one reach with a storage area equal to the storage area of canal reach WM6 and a delay time set at the summation of the delay times of canal reach WM5 and WM6 together. Formula 6.30 presents the dynamic behavior of canal reach WM5 and WM6. The dynamic behavior of all canal reaches is formalized as the Integrator Delay model as described in Paragraph 2.2.2.

$$\Delta e_{WM6}(k+1) = e_{WM6}(k) +$$

$$\Delta e_{WM6}(k) + \frac{T_c}{A_{s,WM6}} \cdot \left(\Delta Q_{c,WM4}\left(k - k_{d,WM5} - k_{d,WM6}\right) - \Delta Q_{c,WM6}(k) - \Delta Q_{L6}(k) \right)$$

Formula 6.30

where k represents the time step index, e_{WM6} the water level deviation from setpoint in canal reach WM6 (m), Δe_{WM6} the change in water level deviation from setpoint in canal reach WM6 (m), $\Delta Q_{c,WM4}$ the change in control flow of gate WM4 (m^3/s), $k_{d,WM5}$ and $k_{d,WM6}$ the number of delay time steps in canal reach WM5 and WM6, respectively, $\Delta Q_{c,WM6}$ the change in control flow of gate WM6 (m^3/s) and ΔQ_{L6} the change in disturbance flow in canal reach WM6. The dynamic behavior of all canal reaches is composed into a large multi-variable state space model. This model is given in Appendix C.
All canal reaches are steep, except for reach WM1. Figure 2.7 reveals that this short, flat, relatively deep reach is sensitive to resonance waves, especially at low flow. The frequency of the basic frequency is estimated by Formulas 2.3 to 2.5 as

3.5e-2 rad/s, which is a period time of 180.2 seconds. These resonance waves can be attenuated by using an appropriate low-pass filter.

6.5.3 Constraints of centralized control of irrigation canal W-M

The constraints applied in the Model Predictive Controller are the minimum and maximum flows of the control structures. The setup of the internal model as the Integrator Delay model with a flow controller to set the gate height, results in a linear internal model. The consequence is that the constraints on the control flow become non-linear. The control flow is a non-linear function of the gate height and the upstream water level as described in the undershot gate Formula 2.9 and, in case the upstream water level is higher than the crest level of the weirs, also of the upstream water level as described in the overshot gate Formula 2.7. The non-linearities are converted into time-variant constraints over the prediction horizon, by taking the previous solution of the water levels as calculated in the optimization of the previous time step. The water levels at the last step of the prediction horizon, which are not calculated in the previous optimization, are set to be equal to the last known water levels in the prediction horizon.
When the upstream water level is lower than the crest of the weirs, the minimum flow constraint is equal to zero. If the upstream water level is higher than the crest of the weirs, the minimum flow constraint is equal to the overshot flow of these weirs. The Model Predictive Controller is not allowed to calculate a control flow that is lower than this overshot flow, as this flow will always occur, even if the undershot gate is set fully closed by the controller.
The maximum flow constraint is equal to the structure flow that occurs when the undershot gate is fully opened by the controller. A fully opened gate must be set in such a way that the bottom of the gate touches the upstream water level. If the gate is lifted out of the water, there will be extra delay when, at a later moment in time, the gate has to be steered back down to decrease the flow again. The water level that develops above the crest of the undershot gate, when the gate is raised, is 2/3rd of the upstream water level above the crest (Brouwer (2004)). This is based on the assumption that super-critical flow can develop over the crest, as with a broad crested weir. So, the maximum flow constraint is calculated from the structure flow, when the gate height is considered to be positioned at 2/3rd of the upstream water level above the crest of the undershot gate. Figure 6.60 presents an example of the open loop water level solution in canal reach WM7 with the resulting minimum and maximum gate position and minimum and maximum flow constraint for gate WM7.

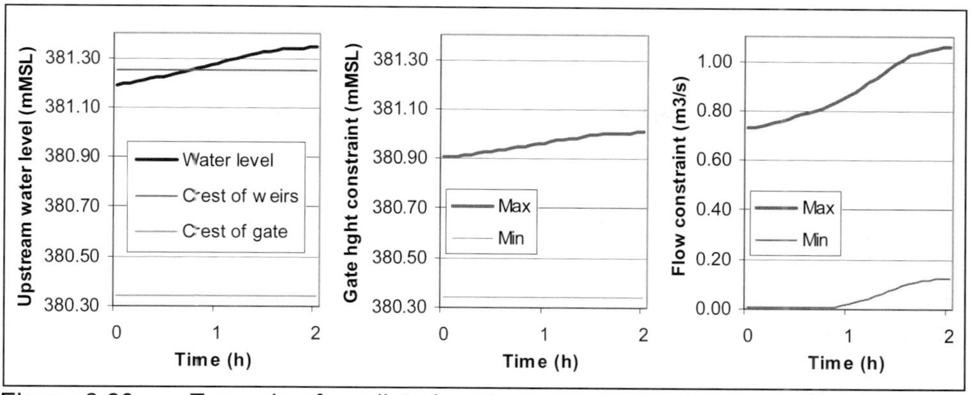

Figure 6.60 Example of predicted upstream water level and resulting time-variant constraints

To convert the change in control flow ΔQ_c that is used as input signal, to an input constraint on the control flow Q_c, the control flow at the previous step is included in the constraint equation:

$$\begin{aligned} \Delta Q(k) &\leq Q_{\max}(k) - Q(k-1) \\ -\Delta Q(k) &\leq -(Q_{\min}(k) - Q(k-1)) \end{aligned}$$ Formula 6.31

These constraints are implemented in all control structures.

6.5.4 Results of centralized control of irrigation canal W-M

The Model Predictive Controller is implemented in a centralized controller and applied to the actual W-M canal. The controller is programmed in C-code (Visual C++ (2000)). The control time step is set at 4 minutes. This results in a prediction horizon of 31 steps given the required 2 hours prediction time horizon. By using this control time step, the computational time for the optimization algorithm is approximately 10 seconds. The cycle of receiving the water level measurements, conditioning the signals, running MPC, sending back the control actions and adjusting the gates, takes about 40 seconds in total. The water levels are strongly filtered before they are used in the controller. The cut-off frequency of the discrete low-pass filter is 2.0e-3 rad/s. The anti-aliasing filtering requires a cut-off frequency of 1.3e-2 rad/s, so the filtering that is actually applied, is stronger than required for this anti-aliasing filtering. An advantage is that the filter strongly attenuates the resonance waves in canal reach WM1. The extra delay caused by the phase lag of the filtering and the duration of the control cycle, is put into the internal model by rounding the delay steps of all reaches upwards to the next multiple of the control time step. As the test is performed after the growing season, when the demanded irrigation flows are low, the values of the delay times and storage areas are taken from the low flow condition from Table 2.3. Table 6.7 gives the parameters of the tested Model Predictive Controller:

Table 6.7 Parameters of Model Predictive Control W-M canal

Parameter	Value	Unit	Remarks
T_c	240	s	
n	31	Steps	2 hours prediction time horizon, with 4 minutes control time step
Q_e	44	-	From e_{MAVE}=0.15 m (see Par. 3.2 for explanation of *MAVE*)
$Q_{\Delta e}$	4.0e4	-	From Δe_{MAVE}=5.0e-3 m
$R_{\Delta Qc}$	4.4e3	-	From $\Delta Q_{c,MAVE}$=0.015 m^3/s
$k_{d,WM1}$	1	Steps	Rounded upwards from $\tau_{d,WM1}$=0
$k_{d,WM2}$	3	Steps	Rounded upwards from $\tau_{d,WM2}$=534
$k_{d,WM3}$	1	Steps	Rounded upwards from $\tau_{d,WM3}$=120
$k_{d,WM4}$	1	Steps	Rounded upwards from $\tau_{d,WM4}$=162
$k_{d,WM5}+k_{d,WM6}$	9	Steps	Rounded upwards from $\tau_{d,WM5}+\tau_{d,WM6}$=1944
$k_{d,WM7}$	3	Steps	Rounded upwards from $\tau_{d,WM7}$=540
$k_{d,WM8}$	5	Steps	Rounded upwards from $\tau_{d,WM8}$=1008
$A_{s,WM1}$	397	m^2	
$A_{s,WM2}$	653	m^2	
$A_{s,WM3}$	503	m^2	
$A_{s,WM4}$	1530	m^2	
$A_{s,WM6}$	1614	m^2	Canal reach WM5 is controlled by local upstream PI-control
$A_{s,WM7}$	2000	m^2	
$A_{s,WM8}$	1241	m^2	
K_p	0.15	-	Local upstream PI-control of gate WM5
K_i	0.05	-	Local upstream PI-control of gate WM5
$T_{c,flowcontrol}$	120	s	Flow controller for all structures, also PI-controlled gate WM5

The resulting state space model has 37 states, 7 inputs and 7 disturbance signals.

During the test only three offtakes are in operation. Two offtakes are constant and the third is increased stepwise twice, and also decreased stepwise twice. One constant offtake and the changing offtake are located in one canal reach, so they are taken together into one summed offtake flow. The intended offtake flows are recorded in a schedule, which is used as feedforward signal in the Model Predictive Controller. Figure 6.61 gives the offtake flows as recorded in the offtake schedule.

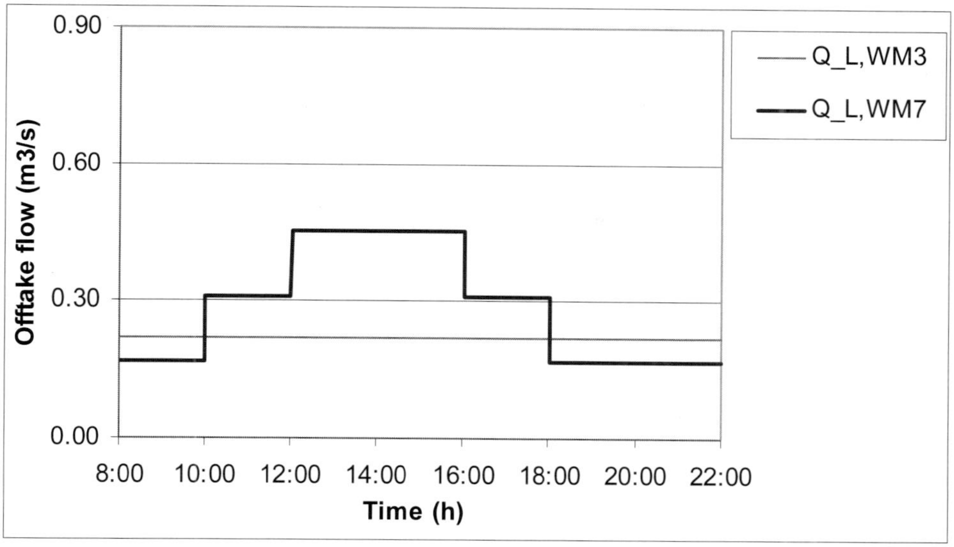

Figure 6.61 Offtake schedule

The control actions to the gates are fully automated in the centralized controller. The offtakes are set manually according to the timing in the offtake schedule. Figures 6.62 to 6.64 give the closed loop results of the test on the actual canal. The heavily oscillating lines in the graphs are caused by the local upstream controlled canal reach WM5. The generated unstable flows are not part of the Model Predictive Controller and do not influence the results of this controller. The downwards spikes are bad values of the data acquisition part. These values are removed manually during the test, by overwriting the value by the previous valid value. The last canal reach WM8 is not in service, so no results of this reach are shown.

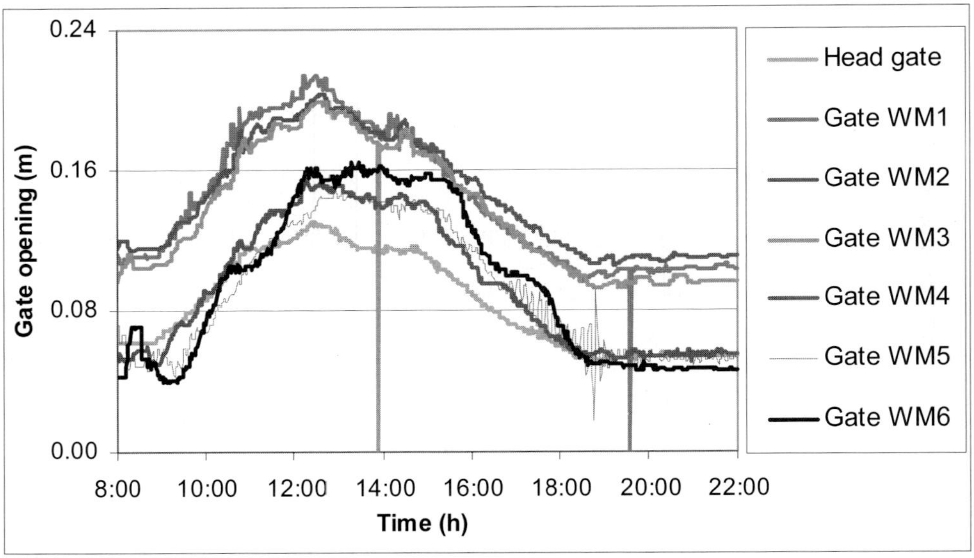

Figure 6.62 Gate opening control gates

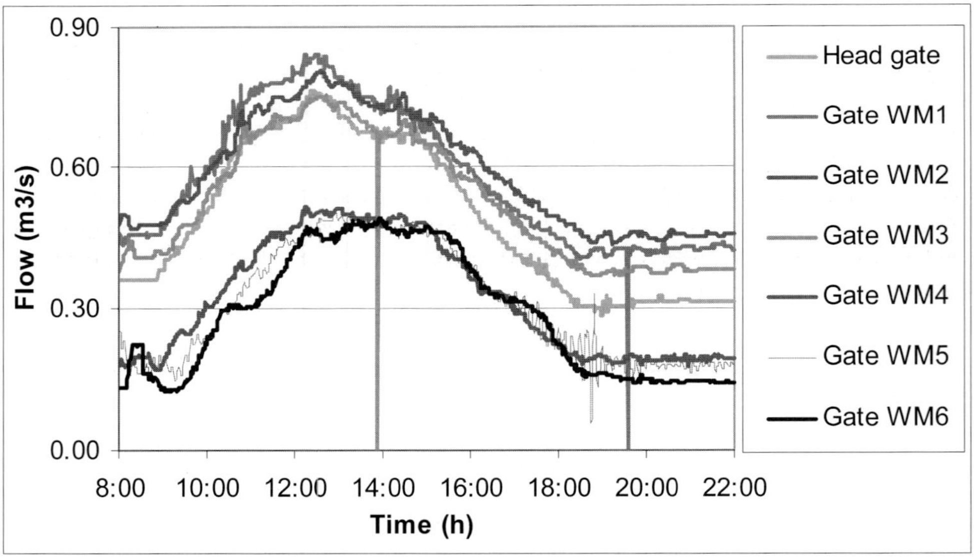

Figure 6.63 Flow control gates

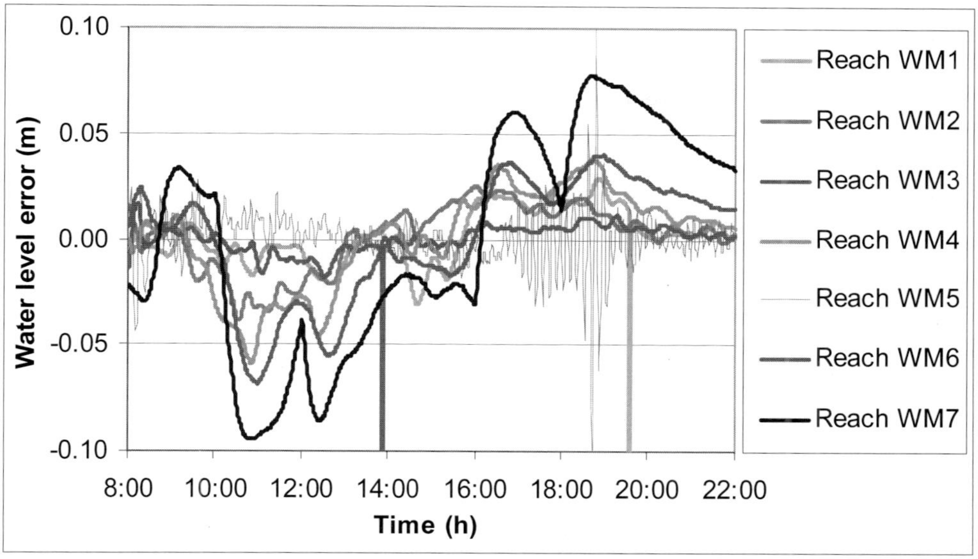

Figure 6.64 Water level deviations from setpoint downstream in the canal reaches

Next, the test is compared to the same test run with the calibrated hydro-dynamic model. Figure 6.65 to 6.71 present the comparison. Exactly the same C-code controller is used, that is linked to the model to calculate the control actions. The gate opening of the head gate differs greatly, as the flow opening of the actual gate is round, while the undershot gate in the model can only be modeled as square. In canal reach WM5, the local upstream PI-controller remains stable in the model. As the oscillations do not disturb the MPC results, no effort is put into either solving the instability in the actual canal or trying to destabilize the local controlled reach in the model.

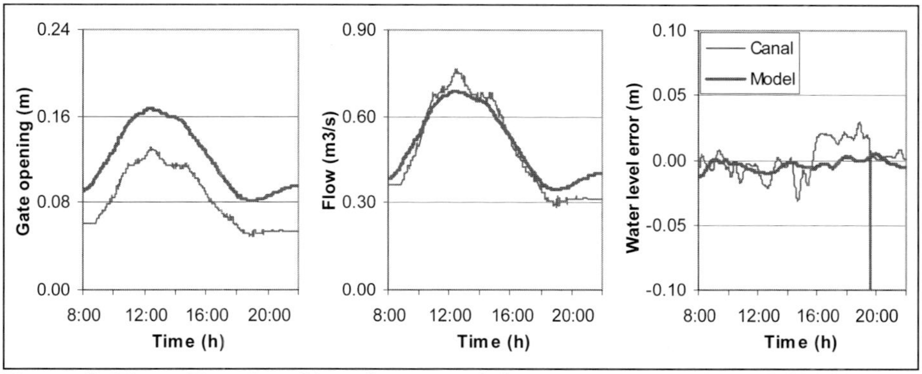

Figure 6.65 Results of head gate and downstream canal reach WM1

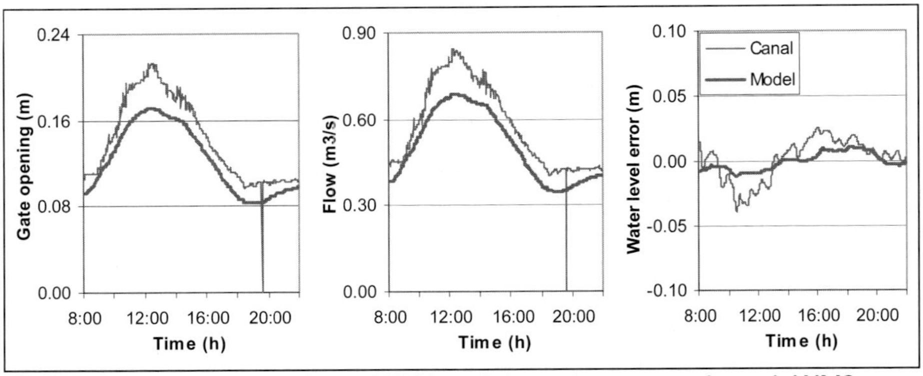

Figure 6.66 Results of gate WM1 and downstream canal reach WM2

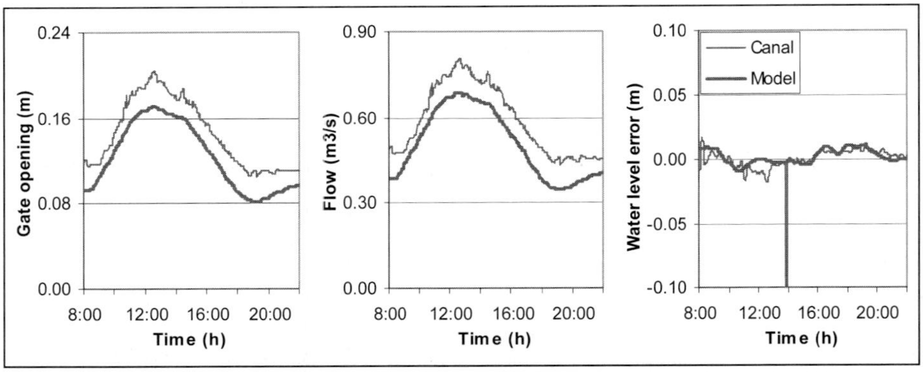

Figure 6.67 Results of gate WM2 and downstream canal reach WM3

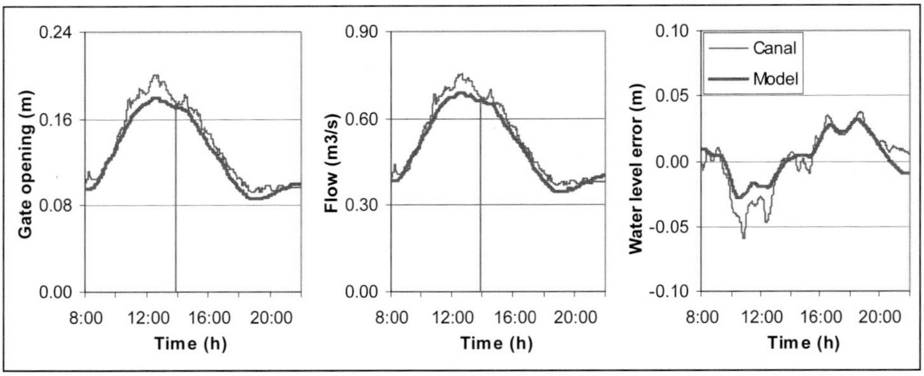

Figure 6.68 Results of gate WM3 and downstream canal reach WM4

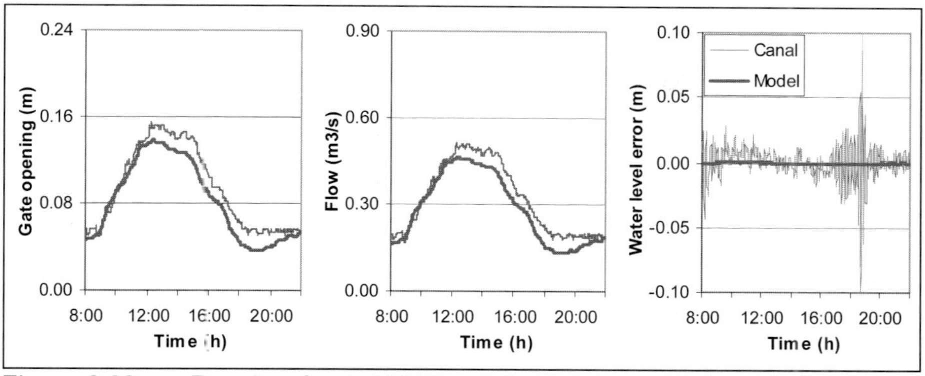

Figure 6.69 Results of gate WM4 and downstream canal reach WM5

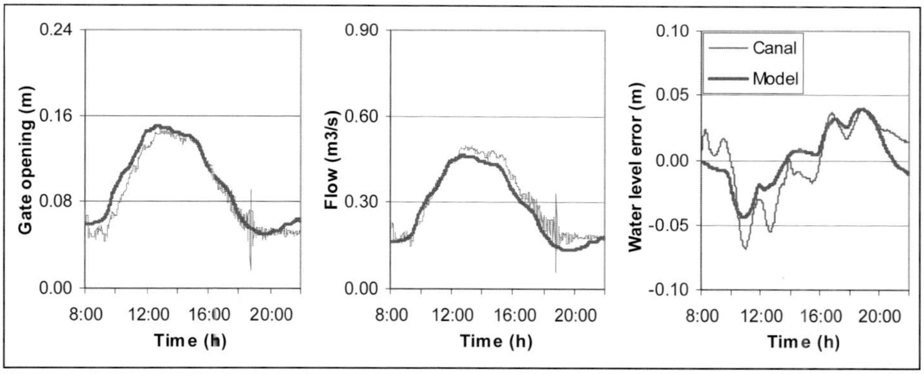

Figure 6.70 Results of gate WM5 and downstream canal reach WM6

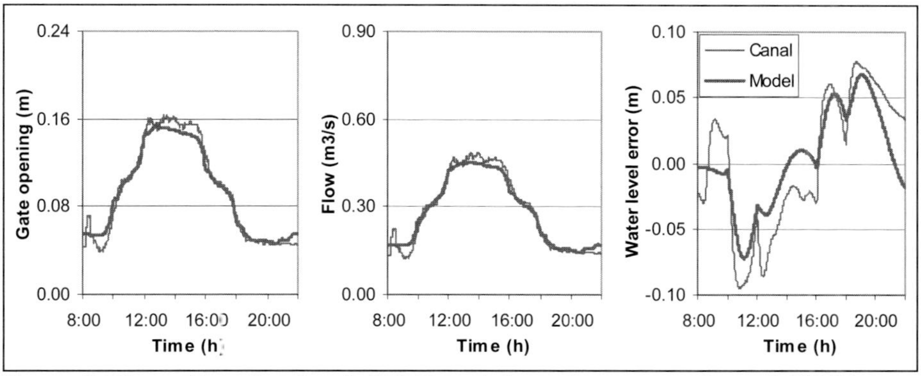

Figure 6.71 Results of gate WM6 and downstream canal reach WM7

6.5.5 Discussions on results of centralized control of irrigation canal W-M

The test is a realistic representation of the regular operation at low flow. The varied offtake flow is located in the last operating canal reach. This means that all control structures and canal reaches are affected by this disturbance. In all canal reaches the water level errors remain within the operating limits approximately 15 centimeters above and below setpoint. All gate openings and gate flows have a smooth course. The controlled water system behaves in a satisfactory manner. This is especially agreeable as the Model Predictive Controller did not need any tuning. It is set according to the Maximum Allowed Value Estimate as explained in Paragraph 3.2. The functioning of the feedforward part of the Model Predictive Controller can be seen in the control actions at all gates that take place before the offtake change is implemented. The head gate starts to release water more than 1 hour before the offtake flow changes at 10:00. This anticipation is visible at all control gates. Even the last gate WM6 releases water a few control time steps before the change in offtake flow.
The differences between the results of the actual test and the test results of the model, are small. The biases in actual and modeled gate opening and actual and modeled flows of the head gate, gate WM1 and gate WM2 are caused by the difference in flow controller of the actual system and the modeled flow controller and unavoidable inaccuracies in the used structure formulas. In the water level results, the same dynamics appear, although the actual canal oscillates more. This is probably due to the way that the structures are modeled. In the model, the structures are described by hydro-static formulas without internal dynamics. In reality, the flow is influenced by the way the flow enters and exits the structure and the pressure waves in the structure. The culverts especially are modeled in a coarse way, which dampens the dynamic behavior.

7 Conclusions and recommendations

In this chapter, the conclusions and recommendations derived from this research are presented.

7.1 Conclusions

The conclusions of this research are categorized according to the chapters in this dissertation. The general conclusions that address the research goals are given in Paragraph 7.1.1. As the applications of Model Predictive Control configurations on various types of open water systems provide foundation for the conclusions on the research goals, the conclusions on the applications are also addressed in this Paragraph 7.1.1. Paragraph 7.1.2 presents the conclusions on the models of open water systems that are implemented in the Model Predictive Controller. In Paragraph 7.1.3, the conclusions on the standard Model Predictive Control configuration, developed in this research are given. In Paragraph 7.1.4, the conclusions on sequential configuration of Model Predictive Control are presented, while the conclusions on multiple model configuration of Model Predictive Control are given in Paragraph 7.1.5.

7.1.1 Conclusions on application of Model Predictive Control in management of open water systems

Water quantity management of open water systems can be formulated as keeping the water levels in the canal reaches close to setpoint, preferably within a limited range around setpoint. This goal is achieved by using as little effort or energy as possible. Disturbance flows in and out of the canals cause the water levels to drift away from setpoint, so the controller has to react on this deviation, comparable to a (disturbance rejecting) feedback controller. The main disturbance flows are usually available for usage as feedforward control. These classical control methods function well in many cases and are compact and easy to understand. These methods should be selected first over more advanced control methods, such as Model Predictive Control. Local feedback control is especially preferable, as communication lines, that are prone to failure, can be avoided. In more and more cases though, the controllability is limited by the constraints on the structure capacities that are used (input constraint) or the narrowing limitation on the water level fluctuations (output constraint). For larger water systems, the management of the total system becomes so complex, that it can only be executed by a very experienced operator or a control method that can incorporate these constraints into the control problem. At present, Model Predictive Control or related constrained model based optimal controllers, are the only control methods that can solve the complex control problem for open water systems taking the constraint into account in a systematic way. In this research, a general Model Predictive Controller for the water quantity management of open water systems is setup. This controller is able to incorporate all water management challenges of irrigation and drainage systems.

Five different types of control on different open water systems are tested, either as Decision Support System for less experienced operators or as fully automated control loop. From these applications, the conclusion can be drawn that Model Predictive Control improves the present water management of open water systems based on classical control methods in case that constraints on structures or on water levels come into play. The improvements are significant when the accuracy of the predictions is reasonable.
The applications are all tuned by using a Maximum Allowed Value Estimate in the weight factors of the variables in the objective function in order to normalize this function. The results of the tests were satisfactory without any further fine-tuning of the weight factors. This proves the general applicability of the Model Predictive Control configuration setup in this research, especially as the method is tested on both drainage systems controlled by large pumps and steep irrigation canals controlled by small gates.
In a design procedure, detailed hydro-dynamic models can be applied to test performance and stability of controllers. This can be concluded from the comparison of the results of a Model Predictive Controller applied to an actual canal and that same controller applied to a hydro-dynamic model. The differences between the results of the two tests are small.
The computational time for the numerical optimization algorithm used in this research, is small enough for real-time implementation of Model Predictive Control on open water systems. This is proven by the implementation of Model Predictive Control on two actual drainage systems and one actual irrigation canal.

7.1.2 Conclusions on modeling of open water systems

The Model Predictive Controller uses models of open water systems to predict future disturbances, states and inputs. The canal reaches of a water system can roughly be categorized as flat, short, deep canals and steep, large, shallow canals. In the first category, the sensitivity to resonance waves is the main characteristic that can destabilize the controlled water system. The canal reaches in this category are referred to as being resonance dominant. The delay time is the main characteristic of the second category that has to be used to achieve a stable controlled water system. These canal reaches are referred to as being delay time dominant. Based on the comparison between the theoretically derived characteristics and the practical identification of these characteristics including the total control loop, it can be concluded that the theoretical values are generally too far off to be of use in the setup of the control. This is because the other parts of the control loop, such as sampling algorithm, filtering, delay time in the communications and structure dynamics, are not taken into consideration.
When the resonance waves have to be controlled or the controller needs to perform well over a wide range of operating points, a De Saint Venant internal model can be used. For delay time dominant canal reaches and for resonance dominant reaches with appropriate filtering, the Integrator Delay model functions well.
From the MPC applications can be concluded that even though the internal models used are simplified to a considerable extent, the controlled water system

performs in a satisfactory manner. This is due to the receding horizon principle in which at every control time step the model is updated with measurements (feedback action) and the latest predictions are acquired (up-to-date feedforward signal). The applications show that the predictions of the future disturbances are accurate enough to be used in model based controllers to improve the water quantity management of open water systems. Although at present, the weak link in Model Predictive Control on drainage systems is the inaccurate forecast of the precipitation.

7.1.3 Conclusions on Model Predictive Control

The Model Predictive Control configuration developed in this research is applicable to all types of open water systems. Application on different systems only requires the use of a different standardized internal model and another set of constraints. The controller can be tuned directly from physical dimensions of the water systems and the requirements on the controlled water system using the Maximum Allowed Value Estimate.
An additional conclusion is that by applying hard constraints on the structures and soft constraints on the minimum and maximum allowed water level, the optimization problem is feasible and convex. This results in a global minimum of the constrained objective function with accompanying optimal control actions.

7.1.4 Conclusions on sequential configuration of Model Predictive Control

By applying Model Predictive Control in a sequential loop, all relevant non-linearities of open water systems can be incorporated in the control problem. In this research, this sequential controller is referred to as sequential configuration of Model Predictive Control.
By using sequential rounding, the off/on constraint that applies to the majority of all pumps can be implemented. In this research, three sequential steps were used to round the off/on pumps. By applying the three sequential steps, off/on pumps are rounded in a more proper way, compared to direct rounding after one optimization step. The sequential rounding results in a lower objective function value with accompanying control actions that have a higher optimum. The consequence of the three sequel steps is an increase in computational time by approximately a factor 3.

7.1.5 Conclusions on multiple model configuration of Model Predictive Control

By applying multiple internal models in parallel each with a certain probability of occurrence, the uncertainty of the disturbance and process model can be quantified. In this way, a stochastic control problem can be solved in a deterministic way by the multiple model configured Model Predictive Controller. The standard objective function is transformed in an objective function, in which the risk is minimized. In this research, risk is defined as the probability of

occurrence of a model times the costs (damage) that results from that model. The multiple models and the accompanying probabilities are derived from a Monte Carlo analysis.
The application of multiple model configuration of MPC on a drainage system with three models, a minimum, average and maximum model, with 10%, 80% and 10% probability, respectively, results in a safer controlled water system. The reason is that in the determination of the control actions, the minimum and maximum scenarios are also weighted and constrained. A disadvantage of using the three models is an increase in computational time of at least a factor 3.

7.2 Recommendations

Model Predictive Control offers a wide applicability in water management of open water systems. On the other hand it requires a central control setup and sufficient computer power. From this research, literature and practical experience in the field of controlled water systems a bottom-up approach is recommended. First, analyze the possibilities of (local) feedback control, next, feedforward control. Only, when constraints limit the proper functioning of the controlled water system, Model Predictive Control is an eminent option. In other words, the selected control method must be as simple as the requirements of the water management, characteristics of the water system and the constraints on the controlled water system allow it to be. To guide this process a procedure is presented that can support the selection of the appropriate control method. The procedure is given in Figure 7.1.

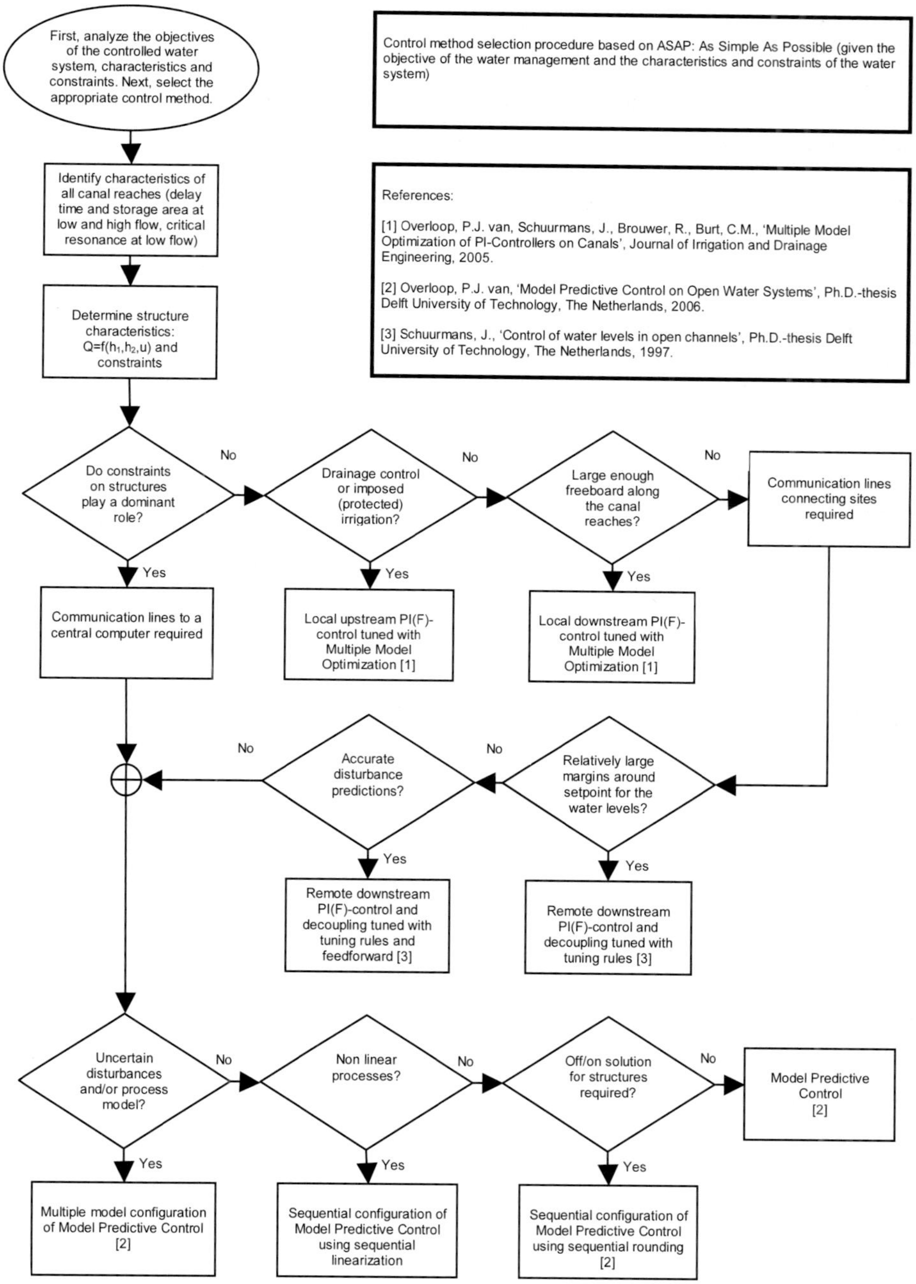

Figure 7.1 Control method selection procedure

As the Model Predictive Control configuration is setup in this research is general and easy to tune, it can be applied to improve the water quantity management of other types of open water systems. Research can be started on the application on water systems with reservoirs that function for flood protection and conservation of water for dry periods. In such a case, long term forecasts of potential evaporation have to be incorporated in the Model Predictive Controller. Another potential application is the real-time determination of the moment at which a flood protection reservoir or an assigned inundation area has to be put into operation. This can be done by sequential rounding in a sequential configured Model Predictive Controller.
As the Model Predictive Controller can incorporate non-linear equations, water quality processes can be incorporated in the controller. By using the Maximum Allowed Value Estimate, the objective function including the water quality variables can be normalized. This enables a direct tuning of the weight factors on these water quality variables. It is suggested to first start testing with the mass conservative substance salt, as saline water intrusion poses an increasing challenge in the water management of low-land areas.
To improve the forecast of the precipitation on drainage systems, radar images for the next hours can be used. This will result in a higher accuracy of the hourly forecast, both in time as well as the location of the rainfall.
The standard Model Predictive Control configuration described in this research is fast enough for real-time implementation on open water systems. For the derived configurations sequential configuration of MPC and multiple model configuration of MPC, faster numerical algorithms or more computer power is required. By applying algorithms that work with sparse matrices, the computational time of the numerical optimization can be decreased, but there is a limit to that decrease. When the total control loop is examined, it can be concluded that the computational load is heavy, but the amount of data that is communicated is small. By using the present state of the communication technology and High Performance Computers at a central location, the computational burden can be minimized. The use of the Internet, Global System for Mobile communication and Short Message Service enables a setup in which locally simple computers acquire, send and receive data over a (international) network and High Performance Computer centers execute the calculations. This setup of the control infrastructure is also interesting for the sustainability of the system, as the Internet and mobile telephony will further develop in an autonomous fashion, also in developing countries and the maintenance on the High Performance Computers and the control software can be done in the clean-rooms of computer centers.

References

Arnold, E., Linke, H., Puta, H. (1998), 'Optimal Control Application for Operational Water Management of a Canal System'. N. T. Koussoulas and P. P. Groumpos, editors, Large Scale Systems: Theory and Applications. 8th IFAC/IFORS/IMACS/IFIP Symposium, Patras, Greece.

Arnold, E., Linke, H., Puta, H. (1999), 'Non-linear Model Predictive Control for Operational Management of a Canal System'. European Control Conference ECC'99, Karlsruhe, Germany.

Bakker, S. (2002), 'Efficient Algorithms for Model Predictive Control'. Literature survey, Delft University of Technology, Department of Mechanical Engineering, System and Control Engineering.

Bautista, A.M., Strelkoff, T.S. & Clemmens, A.J. (2003), 'General Characteristics of Solutions to the Open-Channel Flow, Feedforward Control Problem', Journal of Irrigation and Drainage Engineering, ASCE.

Boom, T.J.J. van & Backx, T.C.P.M. (2001), 'Model Predictive Control', Lecture notes, DISC, Delft, The Netherlands.

Bos, M.G. (1989), 'Discharge measurement structures', ILRI publication 20, The Netherlands.

Bosgra O.H. (2003), 'Introduction Modeling ', Lecture Notes, Delft University of Technology, The Netherlands.

Brouwer R. (2001), 'Operational Water Management', Lecture Notes, Delft University of Technology, The Netherlands.

Brouwer R. (2004), 'Irrigation and Drainage', Lecture Notes, Delft University of Technology, The Netherlands.

Bruggers, M.R. (2004), 'Application Feed Forward Controller on Delta Mendota Canal', Master thesis, Delft University of Technology, The Netherlands.

Burt, C.M. (1999), 'Current Canal Modernization from an International Perspective', Proceedings USCID-Workshop, Phoenix, Arizona.

Burt, C.M. & Piao, X. (2002), 'Advances in PLC-Based Canal Automation', Proceedings USCID-Conference on Benchmarking Irrigation System Performance Using Water Measurement and Water Balances, San Luis Obispo, California.

Burt, C.M., Overloop, P.J. van & Schuurmans, J. (2005), 'Description on How to Tune PI-Controllers for Automatic Upstream and Downstream Control', Report of Irrigation Training and Research Center, San Luis Obispo, California.

Camacho, E.F. & Bordons, C. (1999), 'Model Predictive Control', Springer, New York.

Chow, C.M., Clarke, D.W. (1994), 'Actuator nonlinearities in predictive control', Advanced in Model-Based Predictive Control, Oxford University Press.

Chow, V.T. (1959), 'Open-Channel hydraulics', McGraw-Hill Book Co. Inc, New York.

Clemmens, A.J., Wahl, T.L., Bos, M.G. & Replogle, J.A. (2001), 'Water Measurement with Flumes and Weirs', ILRI publication 58, The Netherlands.

Clemmens, A.J. & Schuurmans, J. (2004), 'Simple Optimal Downstream Feedback Canal Controllers: Theory', Journal of Irrigation and Drainage Engineering, ASCE.

Clemmens, A.J., Strand, R.J., Bautista, E. (2005), 'Field testing of SacMan automated canal control system' Proceedings of USCID Third International Conference on Irrigation and Drainage, San Luis Obispo, California.

Cunge, J.A., Holly, F.M. & Verwey, A. (1980), 'Practical aspects of computational river hydraulics', Boston Pitman.

Dhondia, J., Gaga, M. & Overloop, P.J. van (2000), 'WaterSchedulerOnLine, a DSS-tool for on-line scheduling of water supply', Proceedings International HydroInformatics Conference, Iowa-City.

Eklund, P. & Tufvesson, M. (2001), 'Predictive Control of Irrigation Channels', Master thesis Lund Institute of Technology, Sweden.

Ellerbeck, M.B. (1995), 'Model predictive control for an irrigation canal', Master thesis, Delft University of Technology, The Netherlands.

Gehrels, J.C. (2003), 'Stochastic Hydrology', Lecture Notes, Delft University of Technology, The Netherlands.

Glanzman, G., Siebenthal, M. von, Geyer, T., Papafotiou, G. & Morari, M. (2005), 'Supervisory Water Level Control for Cascaded River Power Plants', Proceedings of Hydropower 05.

Gomez, M., Rodellar, J. & Mantecon, J.A. (2002), 'Predictive Control Method for Decentralized Operation of Irrigation Canals', Applied Mathematical Modeling.

Gonzalez Santos, A. I., Johansen, T. A., Zamarreno Cosme, J. M. (2000), 'Nonlinear Multiple Model Predictive Control in a Fed-batch Reactor', IFAC Symposium on Artificial Intelligence in Real-time Control, Budapest, Hungary.

Hof, A. & Schuurmans, W. (2000), 'Water quality control in open channels', Water Science and Technology.

Holton, G.A. (2004), 'Defining Risk', Financial Analysts Journal, CFA Institute.

Hovland, S. (2004), 'Soft Constraints in Explicit Model Predictive Control', Master thesis, The Norwegian University of Science and Technology, Trondheim, Norway.

Huising, K.C. (2004), 'Slimme Stuwen', Master thesis, Delft University of Technology, The Netherlands.

Kalman, R.E. (1960), 'A New Approach to Linear Filtering and Prediction Problems', Journal of Basic Engineering, ASME.

Kruizinga, S. (2002), 'Statistical Analysis of local 500 hPa ECMWF Ensemble Forecasts', KNMI, The Netherlands.

Kwakernaak, H. & Sivan, R. (1972), 'Linear optimal control systems', John Wiley & Sons, New York.

Leeuwen, A.M. van (2003), 'Voorspellend Sturen van het Boezemstelsel van HHNK', Master thesis, Delft University of Technology, The Netherlands.

Ljung, L. (1987), 'System Identification, theory for the user', New Jersey, PTR Prentice Hall.

Lobbrecht, A.H. (1997), 'Dynamic water-system control', Ph.D. dissertation Delft University of Technology, The Netherlands.

Malaterre, P.O. (1995), 'PILOTE, Optimal Control of Irrigation Canals', International Water Resources Engineering Conference Proceedings, ASCE, New York.

Malaterre, P.O. & Rodellar, J.(1997a), 'Design and comparison of multivariable optimal and predictive control on a 2-pool model', International workshop on regulation of irrigation canals, Marrakech, Morocco.

Malaterre, P.O. & Rodellar, J. (1997b), 'Multivariable predictive control of irrigation canals. Design and evaluation on a 2-pool model', International workshop on regulation of irrigation canals, Marrakech, Morocco.

Malaterre, P.O. & Baume, B.P. (1998a), 'Modeling and regulation of irrigation canals: existing applications and ongoing researches'. Proceedings of IEEE Conference on System, Man and Cybernetics, San Diego.

Malaterre, P.O., Rogers, D.C. & Schuurmans, J. (1998b), 'Classification of Canal Control Algorithms', Journal of Irrigation and Drainage Engineering.

Malda, W. (2005), 'Model Predictive Control on Irrigation Canals, Application on the WM Canal, Arizona', Master thesis, Delft University of Technology, The Netherlands.

Manno, I. (1999), 'Introduction to the Monte Carlo Method', Akadémiai Kiadó, BudaPest, Hungary.

Mareels, I., Weyer, E., Ooi, S.K. & Aughton, D. (2003), 'Modeling and Control of Irrigation Networks: a System Engineering Approach', USCID-Proceedings International Conference, Phoenix.

MathWorks (1992), 'Matlab User Guide', The MathWorks, Inc., Natick, Massachusetts.

Montazar, A. & Overloop, P.J. van (2005), 'Centralized Controller for the Narmada Main Canal', Journal of Irrigation and Drainage.

Mosca, E. (1995), 'Optimal Predictive and Adaptive Control', Prentice Hall, New Yersey.

Murray-Smith, R & Johansen, T.A. (1997), 'Multiple Model Approaches to Modeling and Control', Taylor & Francis, London.

Overloop, P.J. van, Mursi-Batt, A., Heeringen, K.J. van & Thabet, R.A.H.A. (2001), 'Real-Time-Control of Water Quantity and Quality in a Re-Use of Drainage Water Project', Proceedings International ICID-Conference, Seoul, N-Korea.

Overloop, P.J. van (2003a), 'Drainage Control in Water Management of Polders', ICID-proceedings Drainage Workshop 2003, Utrecht, The Netherlands.

Overloop, P.J. van, Schuurmans, W. & Brouwer, R. (2003b), 'Model Predictive Control of water systems in The Netherlands', USCID-Proceedings International Conference, Phoenix.

Overloop, P.J. van (2005a), 'Drainage Control in Water Management of Polders in The Netherlands', Journal of Irrigation and Drainage Systems, 2006.

Overloop, P.J. van, Schuurmans, J., Brouwer, R. & Burt, C.M. (2005b), 'Multiple Model Optimization of PI-Controllers on Canals', Journal of Irrigation and Drainage Engineering.

Overloop, P.J. van, Beuse, P. & Weissenbruch, R. van (2005c), 'Energiezuinig beheer van maal- en spuicomplex IJmuiden', H2O, The Netherlands..

Penning-Rowsell, E.C., Chatterton, J.B. (1977), 'The benefits of flood alleviation: a manual of assessment techniques', Farnborough, Saxon House/Gower, United Kingdom.

Rao,R.R., Aufderheide, B., Bequette, B.W. (1999), 'Multiple Model Predictive Control of Hemodynamic Variables: An Experimental Study', Proceedings of The American Control Conference, San Diego, California.

Rawlings, J.B. (2000), 'Tutorial Overview of Model Predictive Control', IEEE Control Systems Magazine.

Reddy, J.M. (1990), 'Local optimal control of irrigation canals', Journal of Irrigation and Drainage Engineering, ASCE.

Reddy, J. M. (1994), 'Stochastic optimal control of irrigation canals', Advances in Water Resources Technology and Management.

Roos, B. (2003), 'Sturing afvalwaterketen Eindhoven en omgeving', Master thesis, Delft University of Technology, The Netherlands.

Ruiz, V., Schuurmans, J. & Rodellar, J. (1995), 'Properties and Characteristics of Canal Control Algorithms', International Water Resources Engineering Conference Proceedings, ASCE, New York.

Savenije,H.H.G. (2001), 'Hydrology of Catchments, Rivers and Deltas', Delft University of Technology, The Netherlands.

Schuurmans, J. & Ellerbeck, M. (1995a), 'Linear Approximation Model of the WM Canal for Controller Design', International Water Resources Engineering Conference Proceedings, ASCE.

Schuurmans, J. & Liem, G. (1995b), 'Controller design for the WM canal', International Water Resources Engineering Conference Proceedings, ASCE.

Schuurmans, J. (1997), 'Control of water levels in open channels', Ph.D.-dissertation Delft University of Technology, The Netherlands.

Schuurmans, J., Clemmens, A.J., Dijkstra, S.J., Hof, A. & Brouwer, R. (1999a), 'Modeling of Irrigation and Drainage Canals for Controller Design', Journal of Irrigation and Drainage Engineering, ASCE.

Schuurmans, J., Hof, A., Dijkstra, S.J., Bosgra, O.H., Brouwer, R. (1999b), 'Simple Water Level Controller for Irrigation and Drainage Canals', Journal of Irrigation and Drainage Engineering, ASCE.

Schuurmans, W., Grawitz, A.J. & Clemmens, A.J. (1995), 'Guidelines for presentation of Canal Control Algorithms', Proceedings First International Conference Water Resources Engineering, San Antonio.

Schuurmans, W., Hof, A., Overloop, P.J. van, Fish, R. & Hammons, M. (1999), 'Automatic control of the Grand Prairie project', Proceedings USCID-Workshop, Phoenix.

Schuurmans, W., Overloop, P.J. van, Rood, K. & Beukema, P. (2003a), 'Control System for the Bossom Canals in The Netherlands', USCID-Proceedings International Conference, Phoenix.

Schuurmans, W., Overloop, P.J. van & Beukema, P. (2003b), 'Delfland kiest voor volledige automatische bediening van kunstwerken', H2O, The Netherlands..

Seatzu, C. & Usai, G. (2002), 'A decentralized volume variations observer for open channels', Applied Mathematical Modeling.

Singh, V.P. & Woolhiser, D.A., 'Mathematical modeling of watershed hydrology', Journal of Hydrologic Engineering, 2002

Silvis, L.G. (1997), 'Experimental Modeling of Water Movement in Irrigation Canals', Master thesis, Delft University of Technology, The Netherlands.

Silvis, L.G., Hof, A., Hof, P.M.J. & Clemmens, A.J. (1998), 'System identification on open channels', Proceedings 3rd International Conference on Hydroinformatics.

Sobek (2000), 'Manual and Technical Reference', WL|Delft Hydraulics, Delft, The Netherlands.

Sobel' I. M. (1994), 'A primer for the Monte Carlo Method', CRC Press, Boca Raton, Florida.

Steenis, J., Strijker, J., Fermont, A. & Overloop, P.J. van (2003), 'Slimme Stuwen tegen Natte Voeten', H2O, The Netherlands..

Steinbuch, M. (1989), 'Dynamic Modeling and Robust Control of a Wind Energy Conversion System', Ph.D. dissertation Delft University of Technology, The Netherlands.

Stelling, G.S. & Booij, N. (1994), 'Computational Modeling in Open Channel Hydraulics', Lecture Notes, Delft University of Technology, The Netherlands.

Stelling, G.S. & Duinmeyer, S.P.A. (2003), 'A staggered conservative scheme for every Froude number in rapidly varied shallow water flows'. International Journal for Numerical Methods in Fluids.

Stork, M. (2005), 'Model-based optimization of the operational procedure of emulsification', Ph.D.-dissertation Delft University of Technology, The Netherlands.

Tousain, R.L. (2002), 'Dynamic optimization in business-wide process control', Ph.D.-dissertation Delft University of Technology, The Netherlands.

Van Hentenryck, P. (1989), 'Constraint Satisfaction in Logic Programming', MIT Press, Massachusetts.

Vandevegte, J. (1990), 'Feedback control systems', Prentice hall, New Jersey.

Visual C++ (2000), 'User Guide', Compaq Computer Corporation.

Wahlin, B.T. (2002), 'Remote Downstream Feedback Control of Branching Networks', Ph.D. dissertation Arizona State University, Arizona.

Wahlin, B.T. & Bautista, A.M. (2003), 'Feedforward Control: Volume Compensation versus Model Predictive Control', USCID-Proceedings International Conference, Phoenix.

Wahlin, B.T. (2004), 'Performance of Model Predictive Control on ASCE Test Canal 1', Journal of Irrigation and Drainage Engineering, ASCE.

Weijs, S.V. (2004), 'Sturen met onzekere voorspellingen', Master thesis, Delft University of Technology, The Netherlands.

Weissenbruch, R.D. van (2003), 'Onderzoek energieverbruik gemaal IJmuiden', Master thesis, Delft University of Technology, The Netherlands.

Weissenbruch, R.D. van, Overloop, P.J. van & Beuse, P.H. (2004), 'Energieoptimalisatie-module voor spui- en maalcomplex IJmuiden', Eindrapportage Directoraat-Generaal Rijkswaterstaat Regionale Dienst Noord-Holland, The Netherlands.

Weyer, E. (2001), 'System identification of an open water channel', Control engineering practice.

Williamson, D. (1991), 'Digital control and implementation', Prentice Hall, New Jersey.

Wright, S.J. (1997), 'Applying New Optimization Algorithms to Model Predictive Control' Chemical Process Control-V, CACHE, AIChE Symposium Series.

Zagona, E.A. (1992), 'Model predictive control of automated canals', Ph.D. dissertation University of Colorado, Colorado.

List of symbols

Symbol	Description	Unit
A_f	Wetted area of cross section	m^2
A_s	Storage area of canal reach	m^2
$\boldsymbol{A}$	System matrix	
$\boldsymbol{A_c}$	Stabilized system matrix	
$\boldsymbol{B_d}$	Disturbance input matrix	
$\boldsymbol{B_u}$	Control input matrix	
c	Celerity	m/s
C	Chézy friction coefficient	$m^{1/2}/s$
C_g	Calibration coefficient of gate	
$\boldsymbol{C}$	Output matrix	
d	Water depth	m
$\boldsymbol{D}$	Disturbances matrix	
e	Water level deviation from setpoint	m
e_f	Filtered water level deviation from setpoint	m
e^*	Virtual state used in soft constraint	
$\boldsymbol{E}$	States constraint selection matrix	
fu	Linearization factor in discretized Saint Venant equations	
F_c	Filter factor	
$\boldsymbol{f}$	Lagrangian matrix	
$\boldsymbol{F}$	Input constraint selection matrix	
g	Gravitational acceleration	m/s^2
h	Water level	mMSL
h_{cr}	Crest level of gate	mMSL
h_g	Height of gate	mMSL
h_{ref}	Water level setpoint	mMSL
$\boldsymbol{H}$	Hessian matrix	
J	Objective function	
k	Time step index	
k_d	Delay time steps	
K_i	Integral feedback gain factor	
K_p	Proportional feedback gain factor	
$\boldsymbol{K}$	Feedback control matrix	
L	Length of canal reach	m
$\boldsymbol{L}$	State estimator matrix	
M_p	Peak in magnitude of transfer function	
μ	Mean parameter value	
μ_g	Contraction coefficient of gate	
n	Prediction horizon steps	
ξ	Weight factor	
ω	Frequency	rad/s

P_f	Wetted perimeter of cross section	m
P	Probability of occurrence	%
P_e	Power usage of pump	kWatt
q_{lat}	Lateral inflow per unit length along canal reach	$m^3/s/m$
Q	Water flow	m^3/s
Q_c	Control flow	m^3/s
Q_d	Disturbance flow	m^3/s
Q_e	Penalty on state variable e	
$\boldsymbol{Q}$	Weight matrix on states	
ru	Linearization factor in discretized Saint Venant equations	
R_f	Hydraulic radius of cross section	m
R_u	Penalty on input variable u	
$\boldsymbol{R}$	Weight matrix on inputs	
σ	Standard deviation of parameter value	
t	Time	s
T_c	Control time step	s
τ	Delay time	s
u	Input	
u^*	Virtual input used in soft constraint	
$\boldsymbol{U}$	Inputs matrix	
v	Velocity of water flow	m/s
$\boldsymbol{V}$	Left hand side matrix of composed constraints	
W_g	Width of gate	m
$\boldsymbol{w}$	Right hand side matrix of composed constraints	
x	Distance along canal reach	m
$\boldsymbol{X}$	States matrix	
$\boldsymbol{Y}$	Output matrix	
z	Discrete delay step operator	
$\boldsymbol{Z}$	Control actions matrix calculated by Model Predictive Controller	

Appendix A Convex optimization problems

A convex optimization problem is a problem where all of the constraints are convex functions, and the objective is a convex function if minimizing, or a concave function if maximizing. Linear functions are convex, so linear programming problems are convex problems. Conic optimization problems, the natural extension of linear programming problems, are also convex problems.

In a convex optimization problem, the feasible region, the intersection of convex constraint functions, is a convex region, as pictured below.

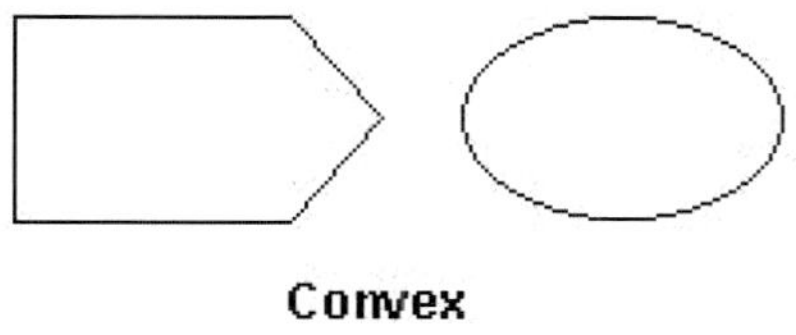

Figure A.1 Convex regions

With a convex objective and a convex feasible region, there can be only one optimal solution, which is globally optimal. Several methods, notably Interior Point methods, will either find the globally optimal solution, or prove that there is no feasible solution to the problem. Convex problems can be solved efficiently up to very large size.

A non-convex optimization problem is any problem where the objective or any of the constraints are non-convex, as pictured below.

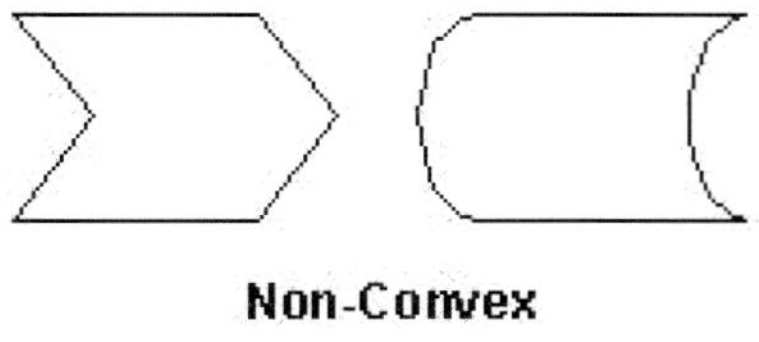

Figure A.2 Non-convex regions

Such a problem may have multiple feasible regions and multiple locally optimal points within each region. It can take time exponential in the number of variables and constraints to determine that a non-convex problem is infeasible, that the objective function is unbounded, or that an optimal solution is the "global optimum" across all feasible regions.

Convex Functions
Geometrically, a function is convex if a line segment drawn from any point (x, f(x)) to another point (y, f(y)), called the chord from x to y, lies on or above the graph of f, as in the picture below:

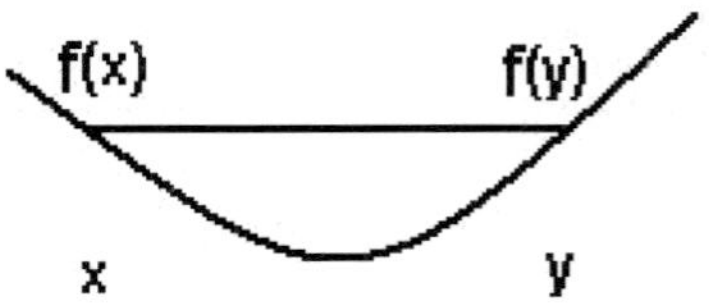

Figure A.3 Convex function

Algebraically, f is convex if, for any x and y, and any t between 0 and 1, f(tx + (1-t)y) <= t f(x) + (1-t) f(y). A function is concave if -f is convex, i.e. if the chord from x to y lies on or below the graph of f. It is easy to see that every linear function, whose graph is a straight line, is both convex and concave.

A non-convex function "curves up and down", it is neither convex nor concave. A familiar example is the sine function:

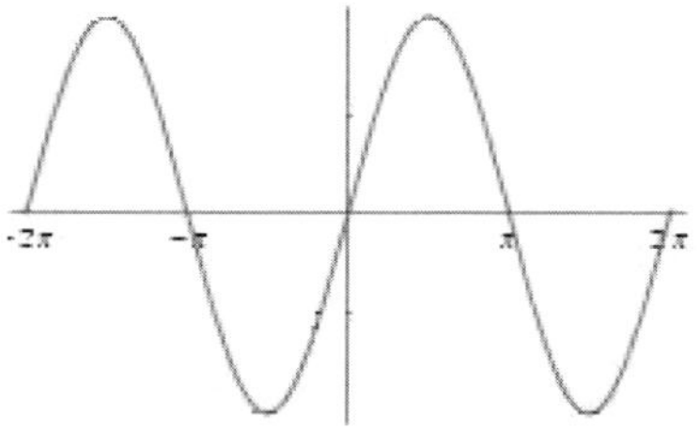

Figure A.4 Non-convex function

but note that this function is convex from -pi to 0, and concave from 0 to +pi. If the bounds on the variables restrict the domain of the objective and constraints to a region where the functions are convex, then the overall problem is convex.

Solving Convex Optimization Problems
Because of their desirable properties, convex optimization problems can be solved with a variety of methods. But Interior Point methods are especially appropriate for convex problems, because they treat linear, quadratic, conic, and smooth nonlinear functions in essentially the same way, they create and use a smooth convex nonlinear barrier function for the constraints, even for LP problems. These methods make it practical to solve convex problems up to very large size, and they are especially effective on second order (quadratic and SOCP) problems, where the Hessians of the problem functions are constant.

Both theoretical results and practical experience show that Interior Point methods require a relatively small number of iterations (typically less than 50) to reach an optimal solution, independent of the number of variables and constraints (though the computational effort per iteration rises with the number of variables and constraints). Interior Point methods have also benefited, more than other methods, from hardware advances, instruction caching, pipelining, and other changes in processor architecture.

Reference: http://www.solver.com/probconvex.htm

Appendix B Quadratic Programming Interior Point algorithm

```
function [z,exitflag] = qpi(H,f,A,b,z0)
%QPI Quadratic programming using Interior Point ALgorithm as described in
% S. Wright, 'Applying New Optimization Algorithms to Model Predictive Control',
% Math. and Comp. Science Division, Argonne Nat. Lab., Argonne, IL 60439
%    X=QPI(H,f,A,b) solves the quadratic programming problem:
%
%             min 0.5*x'Hx + f'x    subject to:  Ax <= b
%              x
%

%initialise
nc = size(A,1);
n2 = size(H,1);
mu = 1;
t = ones(nc,1);
lamda=ones(nc,1);
z=ones(n2,1);
z=z0;
e = ones(nc,1);
sigm=1e-3;
iter = 0;
Dz=1;
exitflag=1;
mu_old=0;
normDzoldold=0;
normDzold=0;
normDz=norm(Dz,inf);

while normDz > 1e-3 & exitflag==1
    rc=H*z+A'*lamda+f';
    rg=-A*z-t+b;
    Lamda = diag(lamda); T=diag(t);
    Tinv = diag(1./t);
    Lamdainv=diag(1./lamda);

    Am = [H+A'*Lamda*Tinv*A];
    bm = -rc+A'*[Lamda*Tinv*rg+lamda-sigm*mu*Tinv*e];
    Dz= Am\bm; % note that inversion only takes place over the 'small' matrix Am!
    Dlamda = -Lamda*Tinv*(-A*Dz + rg + t - sigm*mu*Lamdainv*e);
    Dt = -t + Lamdainv*(sigm*mu*e - T*Dlamda);

    lt = [lamda;t];Dlt = [Dlamda;Dt];

    %----------------------------------------------------------------
    %find alpha_max = sup{alpha in (0,1]| zlt(i)+alpha*Dzlt(i)>0, for i=1,2,..,2*nc}
    %----------------------------------------------------------------
    alpha_m = ones(2*nc,1);
    for i = 1:2*nc
       if Dlt(i)<0 %negative Dzlt(i) gives upper bound on alpha to satisfy condition above
          alpha_m(i,1)=-lt(i)/Dlt(i);
       end
    end
    alpha_max = min(alpha_m);
    %----------------------------------------------------------------

    alpha= min(1,0.995*alpha_max);

    z = z+alpha*Dz;
    lamda = lamda+alpha*Dlamda;
    t = t+alpha*Dt;

    mu = t'*lamda/nc;
    iter = iter+1;
    normDzoldold=normDzold;
    normDzold=normDz;
    normDz=norm(Dz,inf);
    %plot(z);title(['Nr of iterations: ',num2str(iter),'    Stop criterium: ',num2str(normDz),'    mu:
',num2str(mu)]);pause;
    if iter > 40
       %error('The constraints are overly stringent; there is no feasible solution.')
       disp('The constraints are overly stringent; there is no feasible solution.')
       exitflag=-1;
```

```
    end
    if abs(mu-mu_old) < 1e-8 & normDz<normDzold & normDzold<normDzoldold
       disp('The solution is not changing anymore.')
       exitflag=2;
    end
    mu_old=mu;
end
```

Note that in the theory as described in Paragraph 3.4, the matrix *A* and vector *b* as used in this code, are defined as $\boldsymbol{V}$ and $\boldsymbol{w}$, respectively.

Appendix C State space model of W-M canal

The state space model of the W-M canal in Phoenix, Arizona is defined by the state vector x, the system matrix A, the input vector u with the control flows, the input matrix for the control flows B_u, the input vector d with the disturbance flows, the input matrix for the disturbance flows B_d, the output vector y with the water levels at the downstream side of the canal reaches and the output matrix C. The number of states is i. The number of inputs, disturbances and outputs are equal and defined as j.

$$x(k+1) = A \cdot x(k) + B_u \cdot u(k) + B_d \cdot d(k)$$
$$y(k) = C \cdot x(k)$$

where k is the time step index, $x(k) \in \mathbb{R}^{ix1}$, $A \in \mathbb{R}^{ixi}$, $u(k) \in \mathbb{R}^{jx1}$, $B_u \in \mathbb{R}^{ixj}$, $d(k) \in \mathbb{R}^{ixj}$, $B_d \in \mathbb{R}^{ixj}$, $y(k) \in \mathbb{R}^{jx1}$ and $C \in \mathbb{R}^{jxi}$.

$$x(k)=\begin{bmatrix} e_{WM1}(k) \\ \Delta e_{WM1}(k) \\ \Delta Q_{HG}(k-1) \\ e_{WM2}(k) \\ \Delta e_{WM2}(k) \\ \Delta Q_{WM1}(k-1) \\ \Delta Q_{WM1}(k-2) \\ \Delta Q_{WM1}(k-3) \\ e_{WM3}(k) \\ \Delta e_{WM3}(k) \\ \Delta Q_{WM2}(k-1) \\ e_{WM4}(k) \\ \Delta e_{WM4}(k) \\ \Delta Q_{WM3}(k-1) \\ e_{WM6}(k) \\ \Delta e_{WM6}(k) \\ \Delta Q_{WM4}(k-1) \\ \Delta Q_{WM4}(k-2) \\ \vdots \\ \Delta Q_{WM4}(k-9) \\ e_{WM7}(k) \\ \Delta e_{WM7}(k) \\ \Delta Q_{WM6}(k-1) \\ \Delta Q_{WM6}(k-2) \\ \Delta Q_{WM6}(k-3) \\ e_{WM8}(k) \\ \Delta e_{WM8}(k) \\ \Delta Q_{WM7}(k-1) \\ \Delta Q_{WM7}(k-2) \\ \vdots \\ \Delta Q_{WM7}(k-5) \end{bmatrix} \qquad u(k)=\begin{bmatrix} \Delta Q_{HG}(k) \\ \Delta Q_{WM1}(k) \\ \Delta Q_{WM2}(k) \\ \Delta Q_{WM3}(k) \\ \Delta Q_{WM4}(k) \\ \Delta Q_{WM6}(k) \\ \Delta Q_{WM7}(k) \end{bmatrix} \qquad d(k)=\begin{bmatrix} \Delta Q_{L,WM1}(k) \\ \Delta Q_{L,WM2}(k) \\ \Delta Q_{L,WM3}(k) \\ \Delta Q_{L,WM4}(k) \\ \Delta Q_{L,WM6}(k) \\ \Delta Q_{L,WM7}(k) \\ \Delta Q_{L,WM8}(k) \end{bmatrix}$$

The matrices are presented in sparse format. The first number in the location indicator is the row counter, while the second number is the column counter.

$A[1,1]=1$, $A[1,2]=1$, $A[1,3]=\frac{T_c}{A_{s,WM1}}$,

$A[2,2]=1$, $A[2,3]=\frac{T_c}{A_{s,WM1}}$,

$A[4,4]=1$, $A[4,5]=1$, $A[4,8]=\frac{T_c}{A_{s,WM2}}$,

$A[5,5]=1$, $A[5,8]=\frac{T_c}{A_{s,WM2}}$,

$A[7,6]=1$, $A[8,7]=1$, $A[9,9]=1$, $A[9,10]=1$, $A[9,11]=\frac{T_c}{A_{s,WM3}}$,

$A[10,10]=1$, $A[10,11]=\frac{T_c}{A_{s,WM3}}$,

$A[12,12]=1$, $A[12,13]=1$, $A[12,14]=\frac{T_c}{A_{s,WM4}}$,

$A[13,13]=1$, $A[13,14]=\frac{T_c}{A_{s,WM4}}$,

$A[15,15]=1$, $A[15,16]=1$, $A[15,25]=\frac{T_c}{A_{s,WM6}}$,

$A[16,16]=1$, $A[16,25]=\frac{T_c}{A_{s,WM6}}$,

$A[18,17]=1$, $A[19,18]=1$, $A[20,19]=1$, $A[21,20]=1$, $A[22,21]=1$,
$A[23,22]=1$, $A[24,23]=1$, $A[25,24]=1$, $A[26,26]=1$, $A[26,27]=1$,

$A[26,30]=\frac{T_c}{A_{s,WM7}}$,

$A[27,27]=1$, $A[27,30]=\frac{T_c}{A_{s,WM7}}$,

$A[29,28]=1$, $A[30,29]=1$, $A[31,31]=1$, $A[31,32]=1$, $A[31,37]=\frac{T_c}{A_{s,WM8}}$,

$A[32,32]=1$, $A[32,37]=\dfrac{T_c}{A_{s,WM8}}$,

$A[34,33]=1$, $A[35,34]=1$, $A[36,35]=1$, $A[37,36]=1$

The elements in A that have a non-zero value are given in Figure C.1. The other matrices are given in Figure C.2 to C.4.

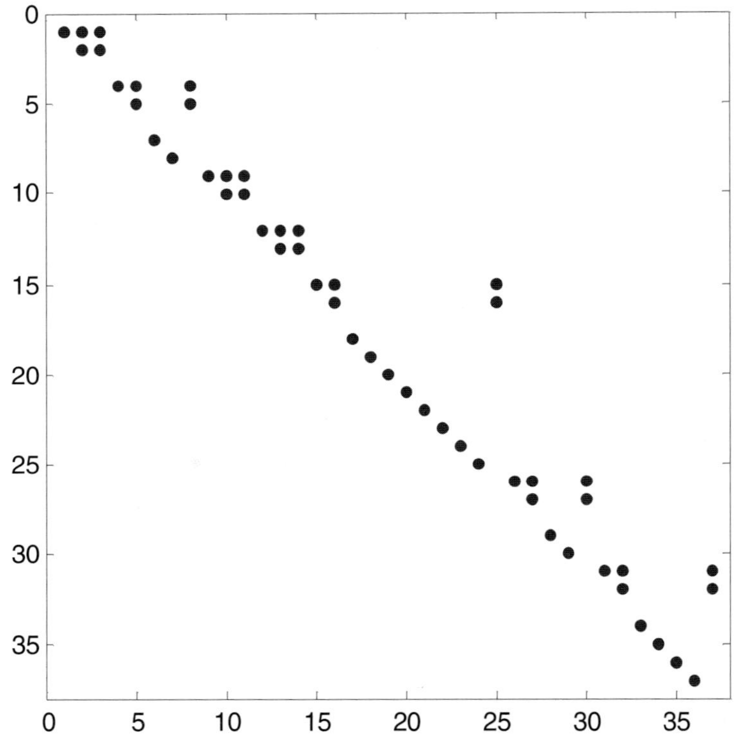

Figure C.1 Non-zero elements in system matrix A, nz = number of non-zero element

$B_u[1,2]=-\dfrac{T_c}{A_{s,WM1}}$, $B_u[2,2]=-\dfrac{T_c}{A_{s,WM1}}$, $B_u[3,1]=1$,

$B_u[4,3]=-\dfrac{T_c}{A_{s,WM2}}$, $B_u[5,3]=-\dfrac{T_c}{A_{s,WM2}}$, $B_u[6,2]=1$,

$$B_u[9,4] = -\frac{T_c}{A_{s,WM3}},\ B_u[10,4] = -\frac{T_c}{A_{s,WM3}},\ B_u[11,3] = 1,$$

$$B_u[12,5] = -\frac{T_c}{A_{s,WM4}},\ B_u[13,5] = -\frac{T_c}{A_{s,WM4}},\ B_u[14,4] = 1,$$

$$B_u[15,6] = -\frac{T_c}{A_{s,WM6}},\ B_u[16,6] = -\frac{T_c}{A_{s,WM6}},\ B_u[17,5] = 1,$$

$$B_u[26,7] = -\frac{T_c}{A_{s,WM7}},\ B_u[27,7] = -\frac{T_c}{A_{s,WM7}},\ B_u[28,6] = 1,$$

$$B_u[33,7] = 1$$

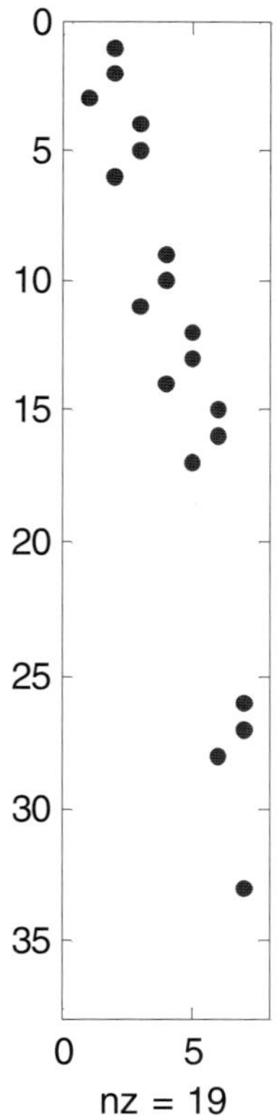

Figure C.2 Non-zero elements in input control matrix B_u

$$B_d[1,1] = -\frac{T_c}{A_{s,WM1}},\ B_d[2,1] = -\frac{T_c}{A_{s,WM1}},$$

$$B_d[4,2] = -\frac{T_c}{A_{s,WM2}},\ B_d[5,2] = -\frac{T_c}{A_{s,WM2}},$$

$$B_d[9,3] = -\frac{T_c}{A_{s,WM3}},\ B_d[10,3] = -\frac{T_c}{A_{s,WM3}},$$

$$B_d[12,4] = -\frac{T_c}{A_{s,WM4}},\ B_d[13,4] = -\frac{T_c}{A_{s,WM4}},$$

$$B_d[15,5] = -\frac{T_c}{A_{s,WM6}},\ B_d[16,5] = -\frac{T_c}{A_{s,WM6}},$$

$$B_d[26,6] = -\frac{T_c}{A_{s,WM7}},\ B_d[27,6] = -\frac{T_c}{A_{s,WM7}},$$

$$B_d[31,7] = -\frac{T_c}{A_{s,WM8}},\ B_d[32,7] = -\frac{T_c}{A_{s,WM8}}$$

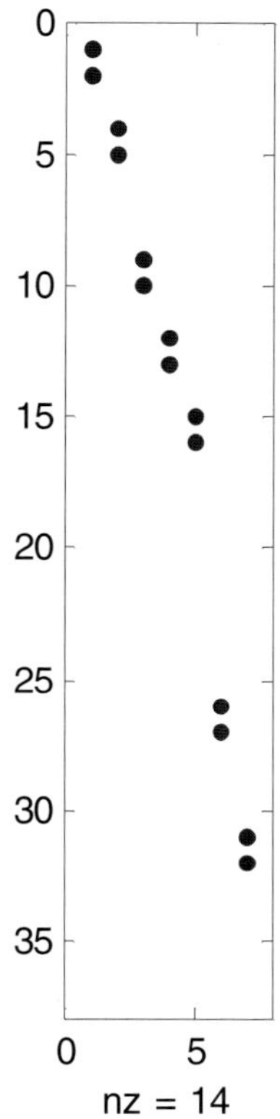

Figure C.3 Non-zero elements in input disturbance matrix B_d

$C[1,1]=1$, $C[2,4]=1$, $C[3,9]=1$, $C[4,12]=1$, $C[5,15]=1$, $C[6,26]=1$, $C[7,31]=1$

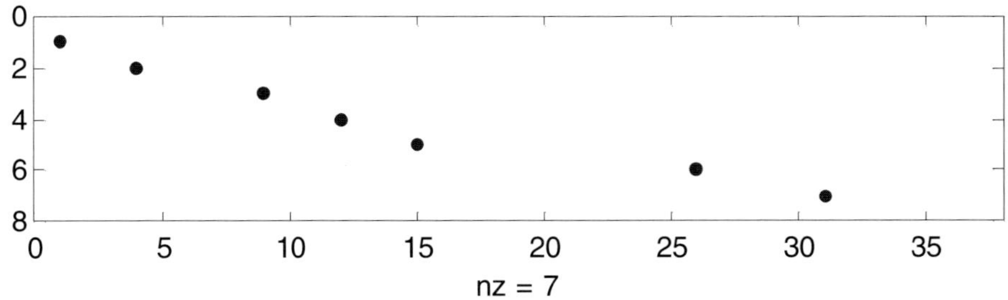

Figure C.4 Non-zero elements in output matrix C

Curriculum Vitae

Name:	Peter-Jules van Overloop
Date of birth:	December 7, 1969
Place of birth:	Bergen op Zoom
Nationality:	The Netherlands

Education:

1982 – 1988	Pre-university education (VWO), Roncalli scholengemeenschap, Bergen op Zoom
1988 – 1994	Delft University of Technology, Mechanical Engineering, mayor in: Systems and Control

Profession:

1994 – 2006	Van Overloop Consultancy. Independent consultant in the fields of modeling, hydro-informatics and control of water systems, hired by institutes, universities and engineering companies in The Netherlands and abroad
2003 – 2004	Nelen en Schuurmans HydroInformatics, Delft. Director and senior consultant
2001 – 2006	Delft University of Technology, Faculty of Civil Engineering and Geosciences, Department of Water Management. Researcher and co-teacher of the course Operational Water Management

Languages:
Peter-Jules van Overloop speaks, reads and writes Dutch and English

Publications (articles on non-water related subjects and professional reports excluded):

Schuurmans, W., Hof, A., Overloop, P.J. van, Fish, R., Hammons, M., 'Automatic control of the Grand Prairie project', Proceedings USCID-Workshop, Phoenix, 1999.

Dhondia, J., Gaga, M., Overloop, P.J. van, 'WaterSchedulerOnLine, a DSS-tool for on-line scheduling of water supply', Proceedings International HydroInformatics Conference, Iowa-City, 2000.

Overloop, P.J. van, Mursi-Batt, A., Heeringen, K.J. van, Thabet, R.A.H.A., 'Real-Time-Control of Water Quantity and Quality in a Re-Use of Drainage Water Project', Proceedings International ICID-Conference, Seoul, N-Korea, 2001.

Overloop, P.J. van, 'Drainage Control in Water Management of Polders', ICID-proceedings Drainage Workshop 2003, Utrecht, The Netherlands, 2003.

Overloop, P.J. van, Schuurmans, W., Brouwer, R., 'Model Predictive Control of water systems in the Netherlands', USCID-Proceedings International Conference, Phoenix, 2003.

Schuurmans, W., Overloop, P.J. van, Beukema, P., 'Delfland kiest voor volledige automatische bediening van kunstwerken', H2O, 2003.

Overloop, P.J. van, Rood, K., Eikelenboom, H., Schuurmans, W. 'Kadebreuk, een Calamiteit?', H2O, 2003.

Schuurmans, W., Overloop, P.J. van, Rood, K., Beukema, P., 'Control System for the Bossom Canals in the Netherlands', USCID-Proceedings International Conference, Phoenix, 2003.

Steenis, J., Strijker, J., Fermont, A., Overloop, P.J. van, 'Slimme Stuwen tegen Natte Voeten', H2O, 2003.

Overloop, P.J. van, Schuurmans, J., Brouwer, R., Burt, C.M., 'Multiple Model Optimization of PI-Controllers on Canals', Journal of Irrigation and Drainage Engineering, 2005.

Montazar, A., Overloop, P.J. van, 'Centralized Controller for the Narmada Main Canal', Journal of Irrigation and Drainage, 2005.

Overloop, P.J. van, Lobbrecht, A., Norel, H. van, Vos, S., 'BOS Hoog Water Groningen', H2O, 2005.

Overloop, P.J. van, Beuse, P., Weissenbruch, R. van, 'Energiezuinig beheer van maal- en spuicomplex IJmuiden', H2O, 2005.

Overloop, P.J. van, Weijs, S.V., Dijkstra, S.J., 'Multiple Model Predictive Control System on Water Systems', Send to Control Engineering Practice, 2006.

Overloop, P.J. van, 'Drainage Control in Water Management of Polders in the Netherlands', Journal of Irrigation and Drainage Systems, 2006.

An up-to-date description of the work of Peter-Jules van Overloop can be found on www.hydraulicscontrol.com and www.voconsult.nl

Samenvatting

De mens is dagelijks afhankelijk van water, met name als drinkwater en voor voedselvoorziening. Ook dient de mens zich te verdedigen tegen een teveel aan water ten gevolge van extreme neerslag of overstromingen. Gemeenschappen van mensen hebben waterbeheerorganisaties gevormd om in deze levensbehoeften te voorzien. Deze organisaties hebben een watersysteem binnen de gemeenschap in beheer en manipuleren de waterstromen in dit systeem om aan de gestelde watergerelateerde doelen te voldoen. Dit wordt gedaan met regelbare kunstwerken, zoals schuiven en pompen. De manier waarop deze kunstwerken worden ingesteld, afhankelijk van de wensen en doelen van de gemeenschappen, betreft het onderzoeksgebied van de meet- en regeltechniek binnen het waterbeheer, ook wel operationeel waterbeheer genaamd.

In het promotieonderzoek 'Model Predictive Control on Open Water Systems' is de relatief nieuwe regelmethodiek Model Predictive Control geconfigureerd voor toepassing op het waterkwantiteitsbeheer van open watersystemen, met name op irrigatiekanalen en grote drainagesystemen. De methodiek maakt gebruik van een intern model van het watersysteem, waarmee optimale regelacties worden berekend over een voorspelhorizon. Voor dit interne model zijn een tweetal vereenvoudigde modellen toegepast; het Integrator Delay model en het Saint Venant model. De initialisatie van de interne modellen vindt plaats door middel van Kalman-filtering. De optimalisatie maakt gebruik van een doelfunctie, waarin conflicterende regeldoelen kunnen worden afgewogen. Deze conflicterende doelen laten zich in de meeste gevallen vertalen in het handhaven van een aantal waterstanden in het watersysteem binnen bepaalde grenzen en het gebruik van een zo minimaal mogelijke regelinspanning of energie daarvoor. Voor het instellen van de weegfactoren in de doelfunctie is in dit onderzoek gebruik gemaakt van een schatting van de maximaal toelaatbare waarde van de variabelen in de doelfunctie. Binnen de uitgevoerde optimalisatie wordt rekening gehouden met de beperkingen die gelden voor de regelbare kunstwerken. De optimalisaties worden iedere regeltijdstap herhaald, waarbij alleen de eerstvolgende reeks van regelacties wordt uitgevoerd. Hierdoor ontstaat een geregeld watersysteem met een continue optimale handhaving van het regeldoel, rekening houdend met verwachtingen, zoals verwachte irrigatiebehoefte of extreme neerslag en met de beperkingen die gelden in het watersysteem.

In dit onderzoek is Model Predictive Control vergeleken met de klassieke regelmethodieken Feedback Control en Feedforward Control om de potentie ervan aan te tonen, vooral voor de problematiek rondom het beheer van watersystemen. Uit deze vergelijking blijkt dat Feedback Control het minst goed presteert in het corrigeren van een afwijking, aangezien er eerst een afwijking moet optreden, voordat de Feedback Controller reageert. Door toevoeging van Feedforward Control wordt een hogere prestatie gehaald. Veel watersystemen worden gekenmerkt door een beperking in de regelmogelijkheden van de kunstwerken; Pompen hebben een gelimiteerde capaciteit en het debiet door schuiven wordt beperkt door de (zee)waterstanden bij het kunstwerk. Door toepassing van Model Predictive Control kunnen deze beperking worden

meegenomen in het berekenen van de regelacties. Hierdoor presteert deze regelmethodiek het best in periodes, waarin de beperking van de kunstwerken een rol spelen. In de overige periodes presteert Model Predictive Control minimaal vergelijkbaar met Feedback Control in combinatie met Feedforward Control. Een ander voordeel van Model Predictive Control is het kunnen omgaan met conflicterende regeldoelen. In de doelfunctie van de optimalisatie worden relatieve weegfactoren toegekend aan de verschillende regeldoelen. Hierdoor wordt een evenwichtige set van regelacties berekend voor het hele watersysteem en kunnen eventuele problemen, zoals een teveel of een tekort aan water, zoveel mogelijk worden uitgespreid over het hele systeem.
De gebruikte optimalisatie is geschikt voor lineaire interne modellen. Aangezien waterstromen in kanalen en kunstwerkdebieten niet-lineair zijn, wordt gebruik gemaakt van het stapsgewijs lineariseren van het optimalisatieprobleem. Door toepassing van een aantal iteratiestappen wordt met het stapsgewijs lineariseren, de niet-lineaire oplossing in voldoende mate benaderd. Hierdoor kunnen alle niet-lineaire onderdelen van een watersysteem, zelfs het sterk niet-lineaire gedrag van een uit- en aanschakelende pomp, worden meegenomen in de optimalisatie.
In het onderzoek is een uitbreiding op de standaard Model Predictive Controller toegepast, waarin onzekerheden in verwachtingen en modellen kunnen worden meegenomen. In plaats van het optimaliseren met één model, worden drie parallelle modellen gebruikt in het interne model van het watersysteem. Eén model representeert het gemiddelde, meest waarschijnlijk scenario. De twee overige zijn het minimale en het maximale scenario. Door vermenigvuldiging met de kans van optreden van deze drie modellen, ontstaat een doelfunctie die wordt geminimaliseerd op risico van hoge waterstanden, in plaats van op de (onzekere) hoge waterstanden zelf. Deze stochastische configuratie van Model Predictive Control wordt Multiple Model Predictive Control genoemd.
Model Predictive Control en de afgeleide configuraties zijn toegepast op nauwkeurige modellen van open watersystemen en op werkelijke irrigatie- en drainagesystemen. De resultaten van de testen tonen duidelijk de meerwaarde aan ten opzichte van de klassieke regelmethodieken. Door de generieke opzet van de methodiek, is deze eenvoudig uit te breiden naar tal van aanliggende werkgebieden, zoals waterkwaliteit, water-krachtopwekking en toepassing op andere type watersystemen zoals kleine reservoirs, rioolstelsels en stuwmeren.
Tot slot wordt opgemerkt, dat het van belang is om de oplossing van een regelprobleem zo eenvoudig mogelijk te houden. Hierbij zijn de doelen van het geregelde watersysteem, de watersysteemkarakteristieken en de beperkingen aan de kunstwerken bepalend voor de complexiteit van de toegepaste regelaar. In veel gevallen zijn lokale Feedback Controllers krachtig genoeg. In andere gevallen is de toepassing van het complexere Model Predictive Control onvermijdelijk. Een selectieprocedure voor de meest geschikte regeling is onderdeel van dit onderzoek.